"This outstanding book on Green Initiatives will be of enormous interest to researchers and practitioners, teachers, and learners alike. Its brilliant and nuanced treatment of diverse green interventions and their form, scope, and effectiveness will be a model for future work on the subject".

Arun Agrawal, *Professor at University of Michigan and Member of National Academy of Sciences in the US*

"Analyzing different conservation incentive programs under the umbrella of "green initiatives", this innovative book directs us towards a topic urgently needing scientific rigor: evaluating policy mixes. How to assess spillover interactions when our featured policy intervention is not the only game in town? Which usually it isn't... Using the US and China as examples and zooming in on particular regions and green policies, the authors convincingly show that some programs implemented in parallel exhibit synergies, yet others tradeoffs: careful empirical analysis is needed to assess which policy mixes have desirable effects. This book shows how we can get started!"

Sven Wunder, *Principal Scientist at European Forest Institute (EFI) in Spain and world-class scholar on payments for environmental services*

"The consequences of PES, IPBES, and other conservation and green initiatives worldwide require detailed attention to the land dynamics generated. This work provides a foundation for such assessments, identifying important but under-addressed dimensions of these initiatives important for conservation, biodiversity, land systems, agricultural, and food sciences, among other fields of research and practice".

B.L. Turner II, *Professor at Arizona State University and Member of National Academy of Sciences in the US*

Conservation Effectiveness and Concurrent Green Initiatives

The book examines concurrent green initiatives and their spillover effects on environmental conservation and management to reveal their impact on conservation effectiveness, drawing on a range of international case studies.

Green initiatives are programs, payments, or endeavors that restore, sustain, or improve nature's capacity, with examples including payments for ecosystem services and the development of nature reserves and protected areas. This book explicitly examines concurrent green initiatives, where initiatives overlap either geographically or in terms of recipients of multiple payments. The book provides a detailed analysis of case studies in the USA and China, including the USA-based Conservation Reserve Program and the Environmental Quality Incentives Program, and the China-based Grain-to-Green Program and the Forest Ecological Benefit Compensation Fund. Through this comparison, the book shows the impact of concurrent green initiatives, including additional or unintended benefits for conservation and local communities as well as negative spillover effects. The book complements these case studies by drawing on other global examples ranging in size from local to continental, including planting native trees and shrubs in Australia and green initiatives in the Baltic Sea region. Overall, this book demonstrates the importance of analyzing concurrent green efforts to better understand both the positive and negative impacts to ensure the optimal effectiveness of these policies and programs for conservation and environmental management.

This book will be of great interest to students and scholars of environmental conservation and management, land use, ecosystem services, and environmental policy, as well as policymakers and practitioners working on environmental initiatives and programs.

Li An is a Professor in the Department of Geography and Center for Complex Human-Environment Systems at San Diego State University, San Diego, CA, USA.

Conghe Song is a Professor in the Department of Geography at the University of North Carolina at Chapel Hill, Chapel Hill, NC, USA.

Qi Zhang is a Research Scholar in the Department of Geography at the University of North Carolina at Chapel Hill, Chapel Hill, NC, USA.

Eve Bohnett is a Postdoctoral Researcher in the Department of Geography at San Diego State University, San Diego, CA, USA.

Routledge Studies in Conservation and the Environment

This series includes a wide range of inter-disciplinary approaches to conservation and the environment, integrating perspectives from both social and natural sciences. Topics include, but are not limited to, development, environmental policy and politics, ecosystem change, natural resources (including land, water, oceans and forests), security, wildlife, protected areas, tourism, human-wildlife conflict, agriculture, economics, law and climate change.

Ethics in Biodiversity Conservation
Patrik Baard

Protected Areas and Tourism in Southern Africa
Conservation Goals and Community Livelihoods
Edited by Lesego Senyana Stone, Moren Tibabo Stone, Patricia Kefilwe Mogomotsi and Goemeone E. J. Mogomotsi

Women and Wildlife Trafficking
Participants, Perpetrators and Victims
Edited by Helen U. Agu and Meredith L. Gore

Conservation, Land Conflicts and Sustainable Tourism in Southern Africa
Contemporary Issues and Approaches
Edited by Regis Musavengane and Llewellyn Leonard

Threatened Freshwater Animals of Tropical East Asia
Ecology and Conservation in a Rapidly Changing Environment
David Dudgeon

Conservation Effectiveness and Concurrent Green Initiatives
Li An, Conghe Song, Qi Zhang, and Eve Bohnett

Religion and Nature Conservation
Global Case Studies
Edited by Radhika Borde, Alison A Ormsby, Stephen M Awoyemi, and Andrew G Gosler

For more information about this series, please visit: www.routledge.com/ Routledge-Studies-in-Conservation-and-the-Environment/book-series/RSICE

Conservation Effectiveness and Concurrent Green Initiatives

Li An, Conghe Song, Qi Zhang, and Eve Bohnett

Routledge
Taylor & Francis Group
LONDON AND NEW YORK

earthscan
from Routledge

First published 2023
by Routledge
4 Park Square, Milton Park, Abingdon, Oxon OX14 4RN

and by Routledge
605 Third Avenue, New York, NY 10158

Routledge is an imprint of the Taylor & Francis Group, an informa business

British Library Cataloguing-in-Publication Data
A catalogue record for this book is available from the British Library

Library of Congress Cataloging-in-Publication Data
A catalog record has been requested for this book

ISBN: 978-1-032-26867-5 (hbk)
ISBN: 978-1-032-26870-5 (pbk)
ISBN: 978-1-003-29029-2 (ebk)

DOI: 10.4324/9781003290292

Typeset in Times New Roman
by Deanta Global Publishing Services, Chennai, India

Contents

Figures

Author contributions

Li An contributed to writing Chapters 2, 3, 5, 8, and 9, and overall book editing. Eve Bohnett contributed to writing Chapters 1, 4, and 9 and overall book editing. Conghe Song contributed to writing several sections in Chapters 8 and 9, the editing of Chapter 6, and overall book editing. Qi Zhang contributed to writing Chapters 6 and 7, and overall book editing.

About the authors

Li An

Center for Complex Human-Environment Systems, San Diego State University; Department of Geography, San Diego State University, San Diego, CA, USA.

Li An is a Professor in the Department of Geography at San Diego State University. He received his Bachelor's degree from Peking University in 1989, Master's degree from the Chinese Academy of Sciences (Systems Ecology; 1992) and Michigan State University (Statistics; 2002), and PhD from Michigan State University in 2003.

Eve Bohnett

Center for Complex Human-Environment Systems, San Diego State University, San Diego, CA, USA.

Eve Bohnett is currently a postdoctoral research fellow at San Diego State University. She earned her PhD (2020) in Landscape Architecture from the University of Florida. She obtained an MSc (2015) in Nature Conservation from Beijing Forestry University and a BSc (2006) in Plant Sciences (Biology) from the University of California, Santa Cruz.

Conghe Song

Department of Geography, University of North Carolina at Chapel Hill, Chapel Hill, NC, USA.

Conghe Song is a Professor of Geography at the University of North Carolina at Chapel Hill. He earned his PhD degree in Geography from Boston University in 2001, Master's degree in Forest Ecology from Beijing Forestry University in 1991, and Bachelor's degree in Forestry from Anhui Agricultural College in 1988.

Qi Zhang

Department of Geography, University of North Carolina at Chapel Hill, Chapel Hill, NC, USA.

Qi Zhang is a Research Scholar at the University of North Carolina (UNC) at Chapel Hill. He earned his PhD degree (2017) and Master's degree (2014) both in Geography from UNC–Chapel Hill. He obtained his Bachelor's degree in Environment and Resources from Zhejiang University (Hangzhou, China) in 2011.

Preface

This book intends to introduce the concepts, practice, and empirical evidence concerning concurrent green initiatives, which are defined to be programs, payments, or endeavors that are made in the same geographic area(s) or involve the same recipient(s) to restore, sustain, or improve nature's capacity to benefit humans. Through extensive data collection and analysis, we discovered that concurrent green initiatives have substantial spillover effects, generating hidden synergies or trade-offs via unexpected pathways. Uncovering and leveraging such hidden spillover effects will help better meet the United Nations' Sustainable Development Goals. Through extensive theoretical and empirical research in the past decade, the authors have identified a strong need: a book addressing concurrent green initiatives and their often-overlooked spillover effects. Under this goal, this book has the following features:

1. This book introduces and provides an overview of essential concepts in nature conservation: green initiatives, payments for environmental services, and concurrent green initiatives. At the same time, this book involves a considerable amount of technical or operational details, including case studies in the USA and China.
2. This book is inspired by two empirical studies funded by the US National Science Foundation, one led by An (http://complexities.org/pes/) and the other by Song (http://csong.web.unc.edu).
3. The findings regarding spillover effects were derived from the two teams' independent data collection and analysis, providing more substantial evidence for such effects.
4. This book highlights the importance of putting ideas or concepts into practice: we post the relevant data, metadata, and models for users' convenience.

Use of this book

Chapters 1 and 2 introduce the background and concepts related to the main topic: green initiatives and spillover effects between concurrent green initiatives. Chapters 3 through 7 present case studies in the USA and China, showing whether spillover effects exist and, if so, in what ways. Chapter 8 presents the evidence of such spillover effects from other places worldwide. Chapter 9 concludes the book with summaries and discussions of relevant issues. Therefore, this book is appropriate for students (graduate students, in particular), faculty, conservation policy analysts, consultants, lawmakers, and government officials who are interested in the main topic and the relevant issues.

This book is also appropriate for senior-level undergraduate and graduate courses related to conservation effectiveness, sustainable development, coupled human and natural systems (CHANS) or socio-ecological systems (SES), payments for environmental (or ecological) services, and the like. A website has been created and maintained to facilitate instructors in teaching and students in better mastering relevant concepts and methods, which can be found at:

http://www.complexities.org/book/green_initiative/

This website aims to provide all datasets (including a metadata file), SAS code, R code, STATA code, and relevant documents. There are three subfolders, EQIP-CRP-data, FNNR-data, and TNNR-data, corresponding to Chapters 3, 5, and 6, respectively.

Acknowledgments

The National Science Foundation partially funded this project under the Dynamics of Coupled Natural and Human Systems program (DEB-1212183, DEB-1313756, and BCS-1826839). This project also received financial support from San Diego State University and the University of North Carolina, Chapel Hill. In particular, the Center for Complex Human-Environment Systems at San Diego State University provides partial funding and space to complete this book project.

We also thank several graduate students: Shuang Yang, Jeanne Patton, Alexandra Yost, Yanjing (Tracy) Liu, Huijie Zhang, and Ren Cao (in the order of time when they were involved in the project). They contributed to this book through collecting data, editing, testing, writing a portion of the code, searching and summarizing literature, producing the graphs, and spotting errors.

1   Background about green initiatives

Humans are degrading or destroying nature's various structures and functions at an alarming rate, jeopardizing essential goods and services upon which the welfare of humanity depends, including food, water, clean air, soil, and biodiversity (Daily & Matson, 2008; Millennium Ecosystem Assessment, 2005). This crisis gave rise to the United Nations' 2030 Sustainable Development Goals in 2015, and Goals 14 and 15 are directly related to conserving life below water and on land (Rosa, 2017). We define green initiatives as any initiatives, including programs, payments, and endeavors, that aim to restore, sustain, or improve nature's capacity to support human well-being. In this context, green initiatives may entail endeavors to sustain or conserve some physical environmental structures and functions, such as the ozone layer in Earth's stratosphere, so that life on Earth's surface is protected from the Sun's ultraviolet radiation (Luken & Grof, 2006). Similarly, green initiatives could be made to protect geological features, landforms, and processes (e.g., glaciers, gushers, and volcano sites) that possess intrinsic, cultural, aesthetic, scientific, or educational value (Kormos et al., 2017).

Under this definition, the United Nations Framework Convention on Climate Change (UNFCCC, 2016), the Green Climate Fund, European Union's Green Deal (European Commission, 2019), and the Reducing Emissions from Deforestation and Forest Degradation (REDD+) program are good examples of green initiatives. Other prominent green initiatives include the so-called payments for environmental services (also named payments for ecological services; PES), programs for integrated conservation and development projects (ICDP), and measures that aim to preserve nature and its services vital to humans. Examples include subsidies, tax exemptions, area-based conservation measures comprised of protected areas, and "other effective area-based conservation measures" (OECMs; Jonas et al., 2014; Maxwell et al., 2020). Green initiatives are widespread across the globe. For instance, the aforementioned REDD+ alone had involved 39 developing countries as of July 2019, covering a forest area of approximately 1.49 billion hectares or 37 percent of the global forest area.

DOI: 10.4324/9781003290292-1

1.1 The concept and popularity of green initiatives

The past two decades have witnessed a large number of initiatives devoted to research on green initiatives, focusing on the principles, design features, implementation, participation and compliance, and socio-environmental impacts and trade-offs (Wunder et al., 2018). A conservative annual monetary value of global environmental services, measured in 2007 $US, was estimated to be US$46 trillion in 1997 and US$145 trillion in 2011 (Costanza et al., 2014), and the latter estimate was twice as much as the global gross national product in the same year (Costanza et al., 1997).

A paper by Ezzine-de-Blas et al. (2016) identified a total of 584 unique payments for environmental (ecosystem) services (PES) programs, among a wide variety of green initiatives worldwide, based on several popular databases (e.g., Science direct and Scopus). The programs were identified using the keywords "payments for environmental services", "payments for ecosystem services", or related terms. Then 55 PES programs were selected, and related program information was collected for the metadata analysis (Table 1.1).

We used these same 55 programs to offer a conservative estimate of the impact of PES programs, which is measured as the total coverage of area and investment. We updated the 55 PES programs and calculated their overall payments based on data and information that became available after 2016 (Table 1.1). Unless there was a one-time buyout format for compensation, the payment for each program was calculated based on the land area involved, compensation rate for a unit area of land, and the program's duration. To make a conservative estimation of the payment amount, we only updated the year of running if we were certain about the program's implementation after 2014 (till 2018). Otherwise, we labeled it as "No" regarding whether it is "still running" in the table.

A few more modifications were applied to Table 1.1, described as follows. (1) For PES programs without "still running" or "year ended" information, we assumed it had run only 2 years to make a very conservative estimate of the total investment; (2) the original paper (Ezzine-de-Blas et al., 2016) mentions that most of the programs were in operation till 2014; therefore the default year of termination was set at 2014 unless we found information indicating otherwise; (3) we searched available information sources mentioned in the paper along with a Google search of all the program names mentioned in Table 1.1 and updated the information about program size, year ended, and payment level. The last column (i.e., References) of Table 1.1 indicates where we obtained updated information for the related PES program: if so, we list the reference; otherwise, we put "S", implying the same as the data in the original paper (Ezzine-de-Blas et al., 2016). When two sources for the same PES program gave different numbers, we chose the one that gave a lower estimate.

Our results indicate that the total land area devoted to these 55 programs is 61.57 million ha, with a total investment of US$140.78 billion. However, China's Natural Forest Conservation Program (NFCP) was not included in the paper (Ezzine-de-Blas et al., 2016). If including forestland areas under NFCP (i.e., 117

Table 1.1 PES program information based on Ezzine-de-Blas et al. (2016)

Id	Id_L	PES program name	Year of start	Program size (ha)	Total payment (USD/ha/year)	Still running	Year ended	# of years	Program overall payment ($)	References*
1	NegB	Los Negros, Bolivia	2003	2,774	1.8	No	2014	11	54,925.2	S
2	PimE	Pimampiro, Ecuador	2000	550	9	No	2014	14	64,400[1]	1
3	ProE	PROFAFOR, Ecuador	1993	22,287	17.1	No	2014	21	7,994,602[1]	2
4	VitF	Vittel (Nestlé Waters), France	1993	5,100	1,283.8	No	2014	21	137,494,980	S
5	SloCh	Sloping Land Conversion, China	1999	33,866,667	263.9	Yes		19	73,028,571,400	3
6	HydoMX	Payments for Hydrological Environmental Services (PSAH), Mexico	2003	598,100	31.9	Yes		15	286,190,850	4
7	ConsUS	Conservation Reserve Program, USA	1985	9,700,000[3]	192[4]	Yes		33	61,459,200,000	5
8	EnvUS	Environmental Quality Incentives Program (EQIP), USA	1996		n.d.	Yes		22		6
9	SenUK	Environmental Quality Incentives Program (EQIP), USA	1987	1,400,000[3]	98.4	No	2017[2]	30	4,132,800,000	7
10	CouUK	Countryside Stewardship Scheme (CSS), United Kingdom	1991	530,620	124.4	Yes		27	1,782,246,456	8
11	NotrG	Northeim model project, Germany	2004	288	208.2	No	2006	2	119,923.2	S
12	CamZ	CAMPFIRE, Zimbabwe	1989	4,300,000	0.2	Yes		29	24,940,000	9

(Continued)

Table 1.1 Continued

Id	Id_L	PES program name	Year of start	Program size (ha)	Total payment (USD/ha/year)	Still running	Year ended	# of years	Program overall payment ($)	References*
13	LavCo	CIPAV-Río La Vieja, Colombia	2003	3,536	18.6	No	2005	2	131,539.2	S
14	ChaE	Chachis, Ecuador	2005	7,200	20	No	2014	9	1,296,000	S
15	ChaCo	Chaina, Colombia	2005	444	328.9	No	2014	9	1,314,284.4	S
16	ProcCo	Procuenca, Chinchina, Colombia	1999	18,650	106.7	No	2014	15	29,849,325	S
17	CelE	Celica, Ecuador	2006	49	96	No	2014	8	37,632	S
18	ChacE	Chaco, Ecuador	2005	70	36	No	2014	9	22,680	S
19	RichS	Richtersveld, South Africa	1991	162,445	0.1	No	2015	24	389,868	10
20	KitK	Kitengela, Kenya	2000	4,650	10[4]	No	2017[2]	17	790,500	11
21	MenM	Menabe, Madagascar	2003		n.d.	No	2008	5	42,500[1]	12
22	TurtT	Sea turtle nest, Tanzania	2001		5.3	No	2014	13		S
23	BirwCa	Bird watch & ecotourism, Cambodia	2004	25,165[3]	0.2	No	2017[2]	13	65,429[1]	13
24	BirnCa	Bird nest protection, Cambodia	2002	200,000[3]	92.5	No	2017[2]	15	277,500,000[1]	14
25	SocE	SocioBosque, Ecuador	2008	1,468,306[3]	30[4]	No	2017[2]	9	396,442,620[1]	15
26	JesH	Jesús de Otoro, Honduras	2002	74	12.4	No	2014	12	11,011.2	S
27	HerCR	Heredia, Costa Rica	2002	2,021	45.4	No	2014	12	1,101,040.8	S
28	SanNC	San Pedro del Norte, Nicaragua	2003	39	26.3	No	2014	11	11,282.7	S
29	KulN	Kulekhani, Nepal	2006	12,492[3]	17.5	No	2014	8	1,748,880	16
30	DonV	Da Nhim PWS, Dong Nai watershed, Vietnam	2009	209,705	15.7	No	2014	5	164,618,42.5	S
31	SonV	Son La PWS, Vietnam	2009	50,900	21	No	2014	5	5,344,500	S

32	OacI	Oach Kalan—Kuhan mini micro watershed, India	2005	10	2.6	No	2014	9	234	S
33	ZapMX	Saltillo, Zapaliname, Mexico	2006	25,000	25[4]	No	2014	8	5,000,000	17
34	SimT	Simanjiro valley, Tanzania	2005	10,000	0.3	No	2014	9	27,000	S
35	BioN	NRCB management, Namibia	1998	6,767,389	0.3	No	2017[2]	19	38,574,117.3	18
36	CatUS	Catskills, NYC, USA	1997	272,592	234	Yes		21	1,339,517,088	19
37	UlgT	Tanzania, PWS	2009	2,240	167.4	No	2016	7	2,624,832	20
38	SilNC	Silvopastoril, Nicaragua	2003	3,139	65.5	No	2005	2	411,209	S
39	SilCR	Silvopastroril, Costa Rica	2003	3,124	77	No	2005	2	481,096	S
40	CidIS	Cidanau watershed PES scheme, Indonesia	2005	100	350[4]	No	2014	9	315,000	21
41	TreU	Uganda, Trees for Global Benefits Programme	2003	1,320	52.1	No	2017[2]	14	962,808	S
42	FirPH	Philippines, No Fire Bonus Scheme	1996	37,500	6.2	No	1998	2	465,000	S
43	TreMZ	Mozambique, carbono.	2002	8,000	60[4]	No	2014	12	5,760,000	22
44	TreMX	Scolel Té	1997	8,947[3]	33.4	Yes		21	6,275,425.8	23
45	MonMX	Monarch	2000	13,551	14.5	No	2014	14	2,750,853	S
46	PwsG	PWS en Munich	1993	1,800	380[4]	No	2014	21	14,364,000	S
47	SumIn	Sumberjaya AF conservation auction	2007	25	171	No	2009	2	8,550	24
48	RefK	Reforestation conservation auction in W Kenya	2009	9	44	No	2011	2	792	S
49	FuqCo	Fuquene, Colombia	2004	202	385.1	No	2014	10	777,902	S
50	KmpfB	Noel Kempff Mercado REDD+ project	1996	642,184[3]	0.6	No	2014	18	6,935,587.2	25
51	PesCR	Payments for Environmental Services (PSA)b, Costa Rica	1997	250,000	71[4]	No	2014	17	301,750,000	26

(Continued)

Table 1.1 Continued

Id	Id_L	PES program name	Year of start	Program size (ha)	Total payment (USD/ha/year)	Still running	Year ended	# of years	Program overall payment ($)	References*
52	MunCo	CIPAV- La Salvajina & PNN Munchique, Colombia	2011	42	1,173.8	No	2014	3	147,898.8	S
53	MakMa	Makira WCS Madagascar	2005	345,764	1	No	2017[2]	12	4,149,168	27
54	LrcB	Landrace conserv payments Bolivia & Peru	2010	8	1,700	No	2012	2	27,200	S
55	BfeB	Bolsa Floresta—Brasil	2010	589,611[3]	60	Yes		8	283,013,280	S

Notes: The information in this table is based on Table 1 in Ezzine-de-Blas et al. (2016), with the updated sources listed at the end of this table. The cells with superscript numbers are the places with updated information based on the source listed in the last column: (1) for overall program payment, (2) for the year ended, (3) for program size, and (4) for total payment.

References of Table 1.1:

1. The original paper (Ezzine-de-Blas et al., 2016).
2. The original paper by Ezzine-de-Blas et al. (2016).
3. Delang, C.O., Yuan, Z. 2015. China's Grain for Green Program. Springer; https://www.nature.com/articles/s41893-017-0004-x.pdf; http://www.chinanews.com/gn/2019/09-05/8948511.shtml.
4. http://pubs.iied.org/pdfs/G04274.pdf.
5. https://www.fsa.usda.gov/programs-and-services/conservation-programs/conservation-reserve-program/;https://www.sciencedirect.com/science/article/pii/S026483 7715002264.
6. https://www.nrcs.usda.gov/wps/portal/nrcs/detail/ms/programs/financial/eqip/?cid=nrcseprd421407.
7. http://jncc.defra.gov.uk/pdf/UKBI2017_F_B1a.pdf.
8. https://www.gov.uk/government/collections/countryside-stewardship-get-paid-for-environmental-land-management; https://www.forestry.gov.uk/countrysidestewardship.
9. http://www.campfirezimbabwe.org/index.php/projects-t/12-community-based-tourism.
10. The original paper (Ezzine-de-Blas et al., 2016).
11. https://kwcakenya.com/wp-content/uploads/2018/01/Report-on-Wildlife-Corridors-and-Dispersal-Areas-Final-July-2017.pdf.
12. The original paper (Ezzine-de-Blas et al., 2016).

13. https://rmportal.net/conservation-enterprises/knowledge-base-page/ce-documents/tmatboey-community-based-ecotourism-project-cambodia/view.
14. https://www.sciencedirect.com/science/article/pii/S0006320712003400.
15. size: http://kenan.ethics.duke.edu/wp-content/uploads/2017/04/6-Ecuador-Socio-Bosque-Presentation-4-12-17.pdf; year: http://kenan.ethics.duke.edu/wp-content/uploads/2017/04/6-Ecuador-Socio-Bosque-Presentation-4-12-17.pdf.
16. http://www.forestaction.org/app/webroot/vendor/tinymce/editor/plugins/filemanager/files/3.%20IASC%20paper%20Khatri.pdf.
17. http://www.watershedmarkets.org/casestudies/Mexico_Zapaliname.html.
18. http://www.apaari.org/web/wp-content/uploads/2018/1st%20IAC%20Proceedings%20and%20Recommendations.pdf.
19. http://www.nyc.gov/html/dep/html/environmental_reviews/upstate_water_supply_resiliency.shtml.
20. The original paper (Ezzine-de-Blas et al., 2016).
21. https://www.eea.europa.eu/atlas/teeb/cidanau-watershed-pes-scheme-indonesia/view.
22. https://search-proquest-com.libproxy.sdsu.edu/docview/1709136707/fulltext/59C6366CEA8B458BPQ/1?accountid=13758.
23. http://www.forestcarbonportal.com/project/scolel-te.
24. https://www.zef.de/fileadmin/webfiles/downloads/projects/auction_workshop/Presentations/Leimona-Beria_ConservationAuction_Indonesia.pdf.
25. https://www.sharp-partnership.org/updates/SHARP_2015_smallholder_PES_review_lores.pdf.
26. https://www.researchgate.net/publication/649_Costa_Rica%27s_Payment_for_Environmental_Services_Program_Intention_Implementation_and_Impact.
27. http://nbsapforum.net/#read-best-practice/994.

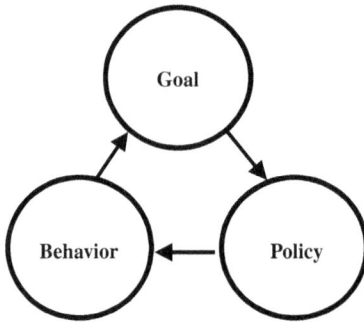

Figure 1.1 Elements of green initiatives. The circles represent the three key elements, and the arrows the influences between elements.

million ha; China Forestry Network, 2013), then the total area of the 56 (55 + 1) PES programs may increase to 178.57 million ha, more than the area of Iran ("World Population Review", 2019).

Regardless of the differences in detail, green initiatives primarily contain the following three elements (Figure 1.1): initiating a *policy*, engaging people in a particular *behavior* or change(s) in *behavior*, and achieving specific *goals* in terms of, e.g., conserving natural capitals, protecting biodiversity, and achieving climate neutrality (Ocasio-Cortez, 2021). The current research focuses on the links between the three elements of the same green initiative: how a certain *goal* may stimulate and formulate a specific *policy*, how a specific *policy* may provide incentives or motivations for people to engage in a particular *behavior* or changes in previous *behaviors*, and whether and how such *behavior* or changes in behaviors may help achieve the *goal*. Undoubtedly, it is imperative to study these elements and the relevant links among them.

1.2 Green initiatives in practice

We begin with introducing several green initiative programs as examples, showing what they are, how they work in real socio-environmental contexts, and what challenges are present in these programs. These programs are selected to cover regions or program types that are not included in the coming chapters.

1.2.1 Incentive-based programs in Nepal

The Chitwan National Park (CNP), Nepal was established in 1973 with the management responsibilities given to the Department of National Parks and Wildlife Conservation. With the Nepali army in charge of enforcing park regulations, resource collection is prohibited in the park except for a 3-day grass gathering window. This policy gave rise to several negative outcomes for local people, which triggered the national park agency and its various partners to develop new complementary policies in the park's buffer zones. Specifically, such policies aim to develop incentive-based programs (IBPs), which can empower local people, e.g., by offering skill training. Furthermore, these policies develop revenue-sharing mechanisms, sustainable extraction regulations, and tourist markets. These programs have a very important goal of continued, sustainable ecosystem protection, which is realized through creating linkages between the social/economic benefits and conservation efforts (Nepal & Spiteri, 2011).

In a survey administered to 189 randomly selected household heads in 2004, questions were asked about their perceptions of program benefits, linkages between conservation and livelihood, and attitudes toward park management. The survey results suggested that because IBPs provide many benefits to local people, it successfully established a perceived direct connection between benefits and livelihoods. About 99% of respondents felt it was good that the land was protected. Tourism turns out to be the primary source of funds, and the continuation of future benefits depends on preserving the park that attracts tourists. Of the survey respondents, 62% reported having obtained benefits from tourism. Of that percentage, 53% had a household member directly employed in tourism services (Nepal & Spiteri, 2011).

However, these policies are reported to have some weaknesses because they are limited in altering the extractive behaviors of local residents. The policies are unable to deliver benefits to the broader population surrounding the park. Villagers far from tourist entry points recognize fewer benefits than the gateway village. Also, extraction opportunities are limited. As a result, the actions of villagers do not always support the views that they express about the importance of conservation. Residents surrounding CNP continue to disregard legal restrictions on resource collection. "Poaching in CNP is often carried out to fulfill subsistence needs of local people, including the collection of forest products for house construction, livestock fodder, and consumption" (Nepal & Spiteri, 2011).

It turns out that IBPs will not guarantee a permanent abandonment of negative behaviors in relation to conservation. Only when benefits outweigh opportunity costs will positive behaviors continue. One major concern is that IBPs may lose persuasion when alternative options that provide greater economic benefits arise. Therefore, "the social and ecological circumstances surrounding CNP suggest IBPs will never preclude the need for effective enforcement mechanisms" (Nepal & Spiteri, 2011).

1.2.2 Multiple green initiatives in Europe

Europe has adopted an integrative conservation approach, which features the High Nature Value farming program. This program aims to connect ecology, land use, and public policies. Woody pastures contribute to landscape-level biodiversity, which simultaneously acts as a repository of genetic resources. This program promotes a range of management practices, including crop rotation, grazing, shrub clearing, and pollarding (a pruning technique), which are instrumental in protecting biodiversity and changing landscape mosaics. Aiming to restore woody pasture landscapes, these practices are widely used as a conservation management method in Western Europe. However, wood-pastures were facing abandonment in recent years (Plieninger et al., 2015).

The Common Agricultural Policy (CAP) provides essential economic support to farmers sustainably managing wood-pastures. The CAP makes direct payments to low-intensity livestock farmers for a variety of ecosystem services they provide. Currently, CAP has established rules that determine which lands and projects are eligible for funding. Member states of the European Union are given the right to determine tree density levels. However, the European Commission can impose heavy fines on member states that are too lax.

Under the European Union's Rural Development Policy, the European Union can make payments to wood-pasture farmers if their lands go above and beyond environmental standards established by CAP. While precise data are not available, few wood-pastures have been involved in this program. Therefore, the support for wood-pastures from the Rural Development Policy could be much more intensive. Additionally, this policy also establishes agro-forestry systems on agricultural land.

A pan-European network of protected areas, known as Natura 2000, is at the core of the European Union's Habitats Directive, which maintains and restores natural habitats. Of the 233 natural habitat types included in this directive, 65 of them have some relationship to wood-pastures even though many are referred to as forest habitats. The criteria for forest habitats under Natura 2000 call for the restoration of tall, ungrazed, dense forests which do not allow sustainable livestock grazing in forests and do not safeguard wood-pastures (Plieninger et al., 2015).

Therefore, there are many policy contradictions surrounding the conservation of European woody pastures. The CAP supports low-intensity farming, while the Rural Development policy seems to supplement those efforts yet also contradicts them by promoting agro-forestry. The Natura 2000 and EU Habitat Directive seem to contradict the CAP.

The study by Pleininger et al. (2015) introduces the CAP as a conservation policy, yet the CAP is more of an overall agriculture policy for the European Union that dates back to the early 1960s. More recent CAP reforms include conservation programs and policies. This is probably why conservation policies within CAP contradict the Habitat Directive and Natura 2000 which were primarily developed for habitat conservation purposes. This issue is probably inherent in other policy contradictions as well. Although programs or policies may consider themselves "green" or "sustainable", their primary goals could be very different or even conflict with one another, which might stall the advancement of other conservation programs or policies.

1.2.3 Green policy-mix in Brazil

The Brazilian Forest Code was established as a federal law, which demands a percentage of rural properties or areas to be maintained as a permanent forest reserve. As of 1996, deforestation was prohibited in 80% of private landholdings in the country's "Legal Amazon" region. However, controversies began to surround this policy. Landowners were not able to easily invest in these areas; for instance, they rarely occupied streambanks with crops and pastures. In parallel to this, governments at various levels did not have the capacity, nor will, to enforce it fully (May et al., 2012).

While the Forest Code continued to maintain the legal baseline, the federal government established an Ecological-Economic Zone (EEZ) in various states within the "Legal Amazon". This policy allocates credit and other public incentives, allowing the reserved area to be reduced to 50% in designated, productive-use areas if they are involved in the EEZ. Forests may be managed for timber and non-timber forest production extraction. Landowners, if not complying with this EEZ policy, must restore forests up to required, baseline limits or purchase "compensation" land elsewhere (May et al., 2012).

Similar to the federal Forest Code, Social, Ecological, and Economic Zoning is implemented as a state zoning strategy. Areas that fall within this policy are part of the PROBIO 2005 listing, which supports Brazil's national policy. This state-level policy also complies with the United Nation's Millennium Development Goals and 2020 Targets adopted in COP10 in Nagoya. These zones have generated controversy, however. Landowners have called for reductions or elimination of the zones. Local, state, and federal authorities have debated microzonation involving local communities.

A complementary policy that aids in conservation is the 1998 Environmental Crime Law. This policy broadens liability for environmental violators, consolidates and imposes greater penalties, and improves the ability of agencies to apply sanctions. At the same time, this policy establishes the liability of corporations and speeds up court proceedings for environmental crimes.

Monitoring of conservation efforts is also enforced at the state level. Since 2000, the State of Mato Grosso has required that all rural properties seek to apply for a license to fell trees, clear brush, and engage in forestry, agriculture, and livestock activities. They must register on an integrated system that relies on satellite images for forest monitoring and inspections with the license. With enrollment waning, a program was launched in 2010 to offer a moratorium on fines in order to stimulate enrollment (May et al., 2012). In addition, a rural credit system complements this program. A direct stimulus for landowners to join the system is the ability to be screened for credit from government banks. Some bank branches require a declaration of compliance with legal reserve requirements and bring properties in line with the Forest Code. Advantageous terms are offered to landowners that are in accordance with environmental codes. Critics of this method claim that environmental restrictions make it more difficult for smallholders to access credit, forcing them to sell their holdings to larger operators (May et al., 2012). Overall, this report portrays a very complementary and synergistic mix of policies at a variety of levels that coordinate to achieve conservation in the Amazon.

1.3 Concurrent green initiatives

The growing impetus to balance ecological and human well-being worldwide has led to simultaneous implementation of multiple green initiatives covering the same geographic area(s) and/or involving the same entities (e.g., persons, households, farms, communities, and groups), which we define as *concurrent* green initiatives. For the popularity of *concurrent* green initiatives, we refer to Section 2.2, where we provide partial evidence that pertains to payments for environmental services only.

Surprisingly, concurrent green initiatives were generally treated as if no-spillover effects (i.e., interrelationships between concurrent green initiatives) existed among them (Figure 1.2), as in the case of the Environmental Quality Incentives Program (EQIP) and the Conservation Reserve Program (CRP)—two of the most extensive concurrent PES programs in the USA (see Chapter 3). Similar to the case in the USA, we have found evidence of spillover effects between two concurrent green initiatives in China's Grain-to-Green Program (GTGP) and Forest Ecological Benefit Compensation (FEBC) Fund. We name this no-spillover presumption, symbolically expressed by the vertical line in Figure 1.2. According to Börner et al.'s synthesis of 14 PES review articles published between 2008 and 2016 (Börner et al., 2017), none explicitly mentioned the existence of concurrent PES programs, let alone examined potential spillover effects among them and the mechanisms underlying such effects. In the very few cases implicitly pointing to interactions between payments (Ezzine-de-Blas et al., 2016; Wunder et al., 2018), very little systematic research—quantitative efforts in particular—has been devoted to revealing and understanding such spillover effects.

Does this no-spillover presumption reflect reality? If not, overlooking these spillover effects may hinder our ability to maximize synergistic effects—or minimize offsetting effects—among concurrent green initiatives. The central goal of this book is thus to determine whether and how such spillover effects exist, understand potential mechanisms behind them, and discuss pathways to escalate the

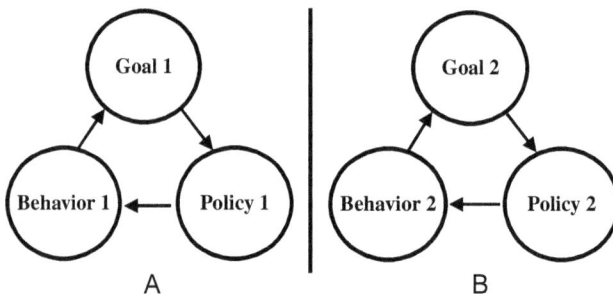

Figure 1.2 Concurrent green initiatives. The vertical line in the middle represents that the two initiatives (one on its left and the other on its right) are treated separately without coordination (i.e., under the no-spillover presumption).

effectiveness of green initiatives. Secondary to this goal, we also aim to show the relevant techniques, models, and methods that are instrumental in detecting the hidden spillover effect between such concurrent green initiatives. This explains why we provide all the data, models (code), and relevant metadata files on the book website (see the Preface of this book).

1.4 A conceptual framework for concurrent green initiatives

Therefore, we design a conceptual framework that examines spillover effects among concurrent green initiatives and integrates the associated elements or links that are displayed in Figure 1.2. We symbolize that Initiatives 1 and 2 stand for two concurrent green initiatives, *Behaviors* 1 and 2 for the (kinds of) behaviors or actions expected from *Policies* 1 and 2, respectively, and *Goals* 1 and 2 for the (kinds of) goals or outcomes expected from *Behaviors* 1 and 2, respectively (Figure 1.3). The current green initiative research broadly focuses on links within the same initiative, specified here as internal links. Specifically, current green initiative literature revolves on (1) whether and in what ways *Policy* 1 may lead

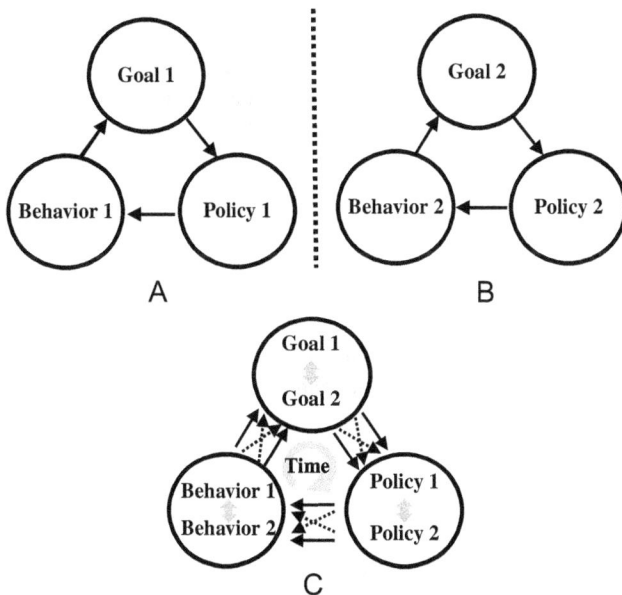

Figure 1.3 Framework for studying concurrent green initiatives. The vertical dotted line stands for removing or minimizing the separation of the two concurrent green initiatives in panels A and B; consequently, we study the concurrent green initiatives as illustrated in panel C. The solid one-way arrows stand for internal influences from one element to another within the same initiative, while the dashed one-way arrows and double two-way arrows for potential spillover effects across different initiatives. The circular one-way arrow represents Time–Time spillover effects.

to *Behavior* 1, generating the *Policy–Behavior* internal link; similarly, current research may explore another internal link from *Policy* 2 to *Behavior* 2, etc.; (2) whether and how *Behavior* 1 may lead to positive (occasionally negative or no) changes in *Goal* 1, such as intended ecosystem services, which represents the *Behavior-Goal* internal link; and in less frequent cases, (3) how changes in *Goal* 1 may loop back and affect *Policy* 1, the original policy scheme, represented as the *Goal–Policy* internal link (Figure 1.3).

What this framework clarifies is the potential spillover effects among concurrent payments (Figure 1.3C). We posit that *Policy* 1 may also affect *Behavior* 2 via *Policy–Behavior* spillover effects. Note that numbers 1 and 2 are interchangeable, as this statement also applies to a spillover effect from *Policy* 2 to *Behavior* 1 (the same hereafter). Similarly, *Behavior* 1 may affect *Goal* 2 via *Behavior-Goal* spillover effects. Equally important is that changes in *Goal* 1 can loop back and affect not only *Policy* 1 via a *Goal–Policy* internal effect (Figure 1.2) but also *Policy* 2 via a *Goal–Policy* spillover effect.

Furthermore, *Policy–Policy* spillover effects may exist, suggesting that one policy directly affects another. To examine such *Policy–Policy* spillover effects, we leverage the current PES stacking literature (for more detail, we refer to Chapter 2). First, horizontal stacking refers to multiple payments made to different parcels of the same recipients, who may respond to these payments differently (e.g., due to time limit) and thus offset or strengthen the goal(s) that would have come out had only one payment been implemented. Second, vertical stacking means that payments are made on the same (or overlapping) parcels, which are often owned or operated by the same recipients. The third type of stacking is temporal stacking, where a landowner receives only one payment at a certain time, yet may receive other payments for different ecosystem services at later times. For more information about these types of stacking, we refer to Section 2.3. Similarly, *Behavior–Behavior* spillover effects and/or *Goal–Goal* spillover effects may occur (Figure 1.3).

In addition to addressing internal links (solid arrows in Figure 1.3), we propose systematically examining spillover effects (dashed arrows in Figure 1.3) and identifying the direction, magnitude, and potential mechanisms behind them. Inspired by the temporal stacking of payments for environmental services, we propose that a temporal dimension (the circular one-way arrow with a Time label; Figure 1.3C) is essential, indicating that internal and spillover effects may evolve.

1.5 The area-based conservation concept

Protected areas serve as the foundation of biodiversity conservation, and more recently, areas outside of protected areas called "other effective area-based conservation measures" (OEABCM) have been recognized for their contribution to nature conservation (Jonas et al., 2014; Maxwell et al., 2020). The Convention on Biological Diversity (CBD) defined area-based conservation measures in 2018 as

> a geographically defined area other than a protected area, which is governed and managed in ways that achieve positive and sustained long-term outcomes

for the *in situ* conservation of biodiversity, with associated ecosystem functions and services and, where applicable, cultural, spiritual, socioeconomic, and other locally relevant values

(CBD, 2018).

In 2018, 193 parties of the CBD adopted 20 "Aichi Targets", including Aichi Target 11 that commits governments to conserve ≥17% of terrestrial and >10% of marine environments globally, especially areas of high conservation importance for biodiversity through securing protected areas or "other area-based conservation areas". Target 11 provides a challenge that will require accelerating designation of appropriate protected areas and securing "other area-based conservation areas" specifically targeted toward biodiversity.

Land-use planning for systematic conservation has begun to incorporate policy tools for sustainable use and restoration of biodiversity and ecosystem services, especially for production landscapes outside of protected areas (Xu et al., 2021). Under OEABCM, programs include areas designated for other management purposes. These areas generate a spectrum of cobenefits, such as supporting the fabric of biodiversity conservation and connectivity on the broader landscape. Examples of areas under the OEABCM concept include places receiving payments for ecosystem services, conservation easements or private conservation lands, military areas, community forests, designated areas with land stewardship implemented by local communities and indigenous peoples, and sustainably managed forestry or fisheries. Deficiencies have been identified in implementing Aichi Targets through inadequate national policy response, funding shortages, science-policy knowledge gaps, and imperfect review mechanisms that can be addressed for more effective implementation of the upcoming post-2020 Global Biodiversity Framework (Xu et al., 2021).

The European Union implemented the European Green Deal to meet the global IUCN Global Standards for Nature-based Solutions. A core part of the European Green Deal includes 2030 Biodiversity Strategies that aim to establish protected areas for 30% of land in Europe and 30% of the sea in Europe. The Green Deal adopts the concept of area-based conservation measures, restoring degraded ecosystems and lands. For instance, the Green Deal promotes sustainable agriculture (e.g., increasing organic farming and biodiversity-rich landscape features on agricultural lands) and aquaculture, reforestation, green infrastructure, river and peatland restoration, and natural coastal protection.

Other conceptual contributions and policy tools have sought to consider meeting specified "retention targets". These targets focus on securing habitat areas that support interconnected biotic and abiotic attributes comprising the corresponding ecosystems (Simmonds et al., 2020). In this way, these areas seek to retain certain species or ecosystem functions as their overarching target. Alternatively, the restoration of degraded land has been the goal of the United Nations' Decade of Ecosystem Restoration program, which aims to restore 350 million acres of land by 2030 and halt and reverse global degradation. However, identifying priority areas on the landscape for restoration becomes challenging in the complexity of ecological, economic, and biocultural considerations. In the USA (Chapter 3) and China (Chapter 4) cases, we design and implement area-based conservation experiments, showing practical ways to maintain the total area of two green initiatives unchanged in the context of budget cuts (more information in Section 1.6).

1.6 The COVID-19-induced budget cuts on green initiatives

The COVID-19 pandemic has placed extensive budget pressure on countries, regions, and organizations worldwide, already or potentially leading to cuts to various sectors, including green initiatives. According to literature, very little of the US$9 trillion of fiscal spending toward the pandemic has been allocated to green policies (Barbier, 2020). The mounting financial burden accruing as part of the COVID-19 pandemic relief efforts has been ravaging economies, which has consequently tied up many financial resources previously anticipated for ensuring progress toward Sustainable Development Goals (SDGs), especially for environmental programs in mitigating climate change, water-related services, and biodiversity (Barbier & Burgess, 2020). Post-pandemic strategies for policy-based environmental programs may involve restructuring distorted policies, which can initiate a transformative reform through approaches such as analysis of policy-mix designs.

Programs like the United Nations' Green Climate Fund (GCF) and the European Union's Green Deal have witnessed the impacts of this crisis on the limited financial resources available to deal with multiple and overlapping issues. These programs have been contending with the high costs of implementing climate adaptation and mitigation actions (Antimiani et al., 2017) and seeking synergies to leverage resources effectively (Markard & Rosenbloom, 2020). These programs require rethinking how transitions toward sustainability can be long-term commitments, requiring policy-mix designs for spending and finance reforms. There are calls for a closer coupling of socioeconomic and environmental goals, seeking to reevaluate policies about air pollution, forestry, and trade in nature-based goods and services to make post-quarantine economic recovery toward a sustainable trajectory (López-Feldman et al., 2020).

A report by the United Nations shows that more than half of the national statistical offices in low-income or lower-middle-income countries encountered COVID-19-related funding constraints. Such constraints are significantly impacting data production for measuring the progress of implementing Sustainable Development Goals (SDGs) (United Nations, 2020, p. 81). In another instance, over 800 research projects worldwide were affected by a significant budget cut to the UK Research & Innovation program, where the reduction in budget ranged from ~US$580 million to ~ US$172 million for the fiscal year of 2021–2022 due to the COVID-19 pandemic (Barclay et al., 2021). The UK's Global Challenges Research Fund experienced similar impacts, and a nearly 50% cut was reported (Foulds et al., 2021).

In the European Union's meeting on July 21, 2020, a budget negotiation reached an agreement on expenditure cuts to essential climate and environmental programs, despite a substantial share of the pandemic recovery budget allocated for climate protection (International Institute for Sustainable Development, 2020). Among the impacted funding pots that carry green goals (e.g., reducing greenhouse gas emissions from burning fossil fuels), InvestEU had a budget cut by 87%, from EUR 31 billion (~US\$37.39 billion) down to less than EUR 4 billion (~US\$4.82 billion), and the Just Transition Fund was cut by more than half from EUR 37.5 billion (~US\$45.23 billion) to EUR 17.5 billion (~US\$21.11 billion) (Reuters News, 2020).

Several Latin American countries have announced a reduction in funds allocated for promoting environmental protection and combating climate change (López-Feldman et al., 2020; Mohan et al., 2021). For instance, 75% of the total budget of the National Natural Protected Area Commission in Mexico was stripped by the federal government (SEGOB, 2020), meaning the cessation of protection for the 25 protected areas regulated by the Yucatán peninsula offices (Varillas, 2020). In El Salvador, with the approved reform of the Budget Law on June 4, 2020, the Ministry of Environment and Natural Resources faced a reduction of US\$1.4 million, a shrinkage of 63% of the resources for protecting natural landscape and wildlife (GatoEncerrado, 2020).

Green funds for protected and conserved areas are severely impacted by the pandemic crisis (IUCN Press, 2021). From January to October 2020, a total of 64 cases from at least 22 countries were found to have experienced rollbacks on green initiatives, especially for environmental protection, which is accompanied by a drastic reduction in budgets for preserved areas (Kroner et al., 2021). A synthesis survey at the global scale showed that half of the protected areas reported cuts in national government funding as a result of the pandemic, and the pandemic most severely impacted conservation efforts in Latin America, Africa, and Asia, with a budget reduction up to 60–70% (Waithaka et al., 2021). According to another survey from 60 countries, 20% of the protected area rangers lost their jobs, and more than 25% experienced salary cuts or delays due to COVID-19-related budget shortage (Singh et al., 2021).

Environmental funding reductions in coping with the pandemic are also observed locally or regionally. For example, the government of Alberta (Canada) decided a province-wide cut of \$5 million (Canadian dollars) of the 2020–2021 budget by entirely or partially closing 20 provincial parks that have critical conservation values for supporting rare local plants and wildlife (CBC News, 2020). The marine preserved area in Nusa Penida (Indonesia) lost 50% of the government funding reallocated to prioritize COVID-19 responses and the loss of tourism revenue (Phua et al., 2021). Facing a budget deficit of nearly US\$9 billion under the COVID-19 crisis, the Parks Department of New York City suffered a cut of US\$84 million (15%) from the US\$540 million budget, threatening the essential services provided by well-managed parks (Columbia Climate School News, 2020; The Hill News, 2020). California ceased the ambitious plan of climate catalyst that aimed to provide a US\$1 billion green loan fund for environmental projects (CalMatters, 2020).

1.7 Summary

Green initiatives are widespread across the world with increasing popularity. In many instances, these initiatives are developed and implemented on a piecemeal basis, lacking overall, comprehensive, and systematic thinking. On the other hand, concurrent green initiatives are becoming prevalent globally, yet very little attention has been given to them. These challenges may give rise to hidden losses or cobenefits in these green initiatives.

Maintaining a considerable amount of protected areas and "other effective area-based conservation measures" is crucially important to conserve nature and vital products and services. This concept accounts for the area-based conservation experiments in Chapters 3 and 4, where we aim to keep constant—at least minimize the loss of—the total area of two or more green initiatives given some level of budget cut. Facing unprecedented challenges such as the COVID-19 pandemic and the subsequent financial crisis, the conservation community should seek to utilize existing funds and resources efficiently. In this regard, we explore a set of measures: integrating and synergizing various policy instruments, minimizing redundancies in green initiative efforts, and exploring pathways that lead to sustainable human–environment dynamics in the following chapters.

References

Antimiani, A., Costantini, V., Markandya, A., Paglialunga, E., & Sforna, G. (2017). The Green Climate Fund as an effective compensatory mechanism in global climate negotiations. *Environmental Science & Policy*, *77*, 49–68. https://doi.org/10.1016/j .envsci.2017.07.015

Barbier, E. B. (2020). Greening the post-pandemic recovery in the G20. *Environmental and Resource Economics*. https://doi.org/10.1007/s10640-020-00437-w

Barbier, E. B., & Burgess, J. C. (2020). Sustainability and development after COVID-19. *World Development*, *135*, 105082. https://doi.org/10.1016/j.worlddev.2020.105082

Barclay, J., Buse, K., Horwell, C. J., & Hawkes, S. (2021). Thousands protest against funding cuts to SDG work. *Nature*, *592*, 353.

Börner, J., Baylis, K., Corbera, E., Ezzine-de-Blas, D., Honey-Rosés, J., Persson, U. M., & Wunder, S. (2017). The Effectiveness of payments for environmental services. *World Development*, *96*, 359–374. https://doi.org/10.1016/j.worlddev.2017.03.020

CalMatters. (2020, May 14). *New state park? Climate catalyst fund? On hold in Newsom's coronavirus-afflicted budget*. https://calmatters.org/environment/2020/05/california -climate-environment-budget-may-revise/

CBC News. (2020, March 3). 20 Alberta parks to be fully or partially closed, while dozens opened up for "partnerships". *CBC News*. CBC. https://www.cbc.ca/news/canada/ calgary/alberta-park-funding-slashed-1.5484095

CBD. (2018). *Convention on biological diversity. COP 14. Decision 14/8. Protected areas and other effective area-based conservation measures*. Convention on Biological Diversity. https://www.cbd.int/decision/cop/?id=13650

China Forestry Network. (2013). *Public report on China's greenness status in 2013*. www .forestry.gov.cn

Columbia Climate School News. (2020, August 31). *The Mayor's shameful mismanagement of New York City's parks.* https://news.climate.columbia.edu/2020/08/31/mayors-shameful-mismanagement-new-york-citys-parks/

Costanza, R., d'Arge, R., de Groot, R., Farber, S., Grasso, M., Hannon, B., Limburg, K., Naeem, S., O'Neill, R. V., Paruelo, J., Raskin, R. G., Sutton, P., & van den Belt, M. (1997). The value of the world's ecosystem services and natural capital. *Nature, 387*(6630), 253–260. https://doi.org/10.1038/387253a0

Costanza, R., de Groot, R., Sutton, P., van der Ploeg, S., Anderson, S. J., Kubiszewski, I., Farber, S., & Turner, R. K. (2014). Changes in the global value of ecosystem services. *Global Environmental Change, 26*, 152–158. https://doi.org/10.1016/j.gloenvcha.2014.04.002

Daily, G. C., & Matson, P. A. (2008). Ecosystem services: From theory to implementation. *Proceedings of the National Academy of Sciences of the United States of America, PNAS, 105*(28), 9455–9456. https://doi.org/10.1073/pnas.0804960105

European Commission. (2019). *The European green deal: Communication from the commission to the European Parliament, the European council, the council, the European economic and social committee and the committee of the regions.* https://ec.europa.eu/info/sites/info/files/european-green-deal-communication_en.pdf

Ezzine-de-Blas, D., Wunder, S., Ruiz-Pére, M., & Moreno-Sanchez, R. del P. (2016). Global patterns in the implementation of payments for environmental services. *PLOS ONE, 11*(3): e0149847.

Foulds, C., Jones, A., & Bharucha, Z. P. (2021). UK aid and research double accounting hits SDG projects. *Nature, 592*(7853), 188.

GatoEncerrado. (2020, June 5). *Gobierno reduce presupuesto a prevención de riesgo en medio de emergencia.* GatoEncerrado. https://gatoencerrado.news/2020/06/05/gobierno-reduce-presupuesto-a-prevencion-de-riesgo-en-medio-de-emergencia/

International Institute for Sustainable Development. (2020, July 22). *New EU budget and recovery fund: Green stimulus and climate budget cuts.* https://www.iisd.org/sustainable-recovery/news/new-eu-budget-and-recovery-fund-green-stimulus-and-climate-budget-cuts/

IUCN Press. (2021, March 11). *COVID-19 fallout undermining nature conservation efforts—IUCN publication.* https://www.iucn.org/news/world-commission-protected-areas/202103/covid-19-fallout-undermining-nature-conservation-efforts-iucn-publication

Jonas, H. D., Barbuto, V., Jonas, H. C., Kothari, A., & Nelson, F. (2014). New steps of change: Looking beyond protected areas to consider other effective area-based conservation measures. *PARKS, 20*(2), 111–128. https://doi.org/10.2305/IUCN.CH.2014.PARKS-20-2.HDJ.en

Kormos, C. F., Badman, T., Jaeger, T., Bertzky, B., van Merm, R., Osipova, E., Shi, Y., & Larsen, P. B. (2017). *World heritage, wilderness, and large landscapes and seascapes* (No. 978-2-8317-1847-7). IUCN. https://www.researchgate.net/profile/James_Allan14/publication/318106483_Chapter_4_Current_wilderness_coverage_on_the_World_Heritage_List_Broad_gaps_and_opportunities/links/5959995b0f7e9ba95e12677c/Chapter-4-Current-wilderness-coverage-on-the-World-Heritage-List-Broad-gaps-and-opportunities.pdf

Kroner, R. G., Chassot, O., Chaudhary, S., Cordova jr, L., Trinidad, A. C., Cumming, T., Howard, J., Said, C. K., Kun, Z., Ogena, A., Palla, F., Valiente, R. S., Troëng, S., Valverde, A., Wijethunga, R., & Wong, M. (2021). COVID-era policies and economic recovery plans: Are governments building back better for protected and conserved areas? *Parks, 27*, 135–148.

López-Feldman, A., Chávez, C., Vélez, M. A., Bejarano, H., Chimeli, A. B., Féres, J., Robalino, J., Salcedo, R., & Viteri, C. (2020). Environmental impacts and policy responses to Covid-19: A view from Latin America. *Environmental and Resource Economics*, 1–6. https://doi.org/10.1007/s10640-020-00460-x

Luken, R., & Grof, T. (2006). The Montreal protocol's multilateral fund and sustainable development. *Ecological Economics*, *56*(2), 241–255. https://doi.org/10.1016/j.ecolecon.2004.04.013

Markard, J., & Rosenbloom, D. (2020). A tale of two crises: COVID-19 and climate. *Sustainability: Science, Practice and Policy*, *16*(1), 53–60. https://doi.org/10.1080/15487733.2020.1765679

Maxwell, S. L., Cazalis, V., Dudley, N., Hoffmann, M., Rodrigues, A. S. L., Stolton, S., Visconti, P., Woodley, S., Kingston, N., Lewis, E., Maron, M., Strassburg, B. B. N., Wenger, A., Jonas, H. D., Venter, O., & Watson, J. E. M. (2020). Area-based conservation in the twenty-first century. *Nature*, *586*(7828), 217–227. https://doi.org/10.1038/s41586-020-2773-z

May, P., Andrade, J., Vivan, J., Kaechele, K., Gebara, M. F., & Abad, R. (2012). *Assessment of the role of economic and regulatory instruments in the conservation policymix for the Brazilian Amazon: A coarse grain analysis.*

Millennium Ecosystem Assessment. (2005). *Ecosystems and Human Well-being*. Island Press.

Mohan, M., Rue, H. A., Bajaj, S., Galgamuwa, G. A. P., Adrah, E., Aghai, M. M., Broadbent, E. N., Khadamkar, O., Sasmito, S. D., Roise, J., Doaemo, W., & Cardil, A. (2021). Afforestation, reforestation and new challenges from COVID-19: Thirty-three recommendations to support civil society organizations (CSOs). *Journal of Environmental Management*, *287*, 112277. https://doi.org/10.1016/j.jenvman.2021.112277

Nepal, S., & Spiteri, A. (2011). Linking livelihoods and conservation: An examination of local residents' perceived linkages between conservation and livelihood benefits around Nepal's Chitwan National Park. *Environmental Management*, *47*(5), 727–738. https://doi.org/10.1007/s00267-011-9631-6

Ocasio-Cortez, A. (2021). Recognizing the duty of the Federal Government to create a Green New Deal. https://www.congress.gov/bill/117th-congress/house-resolution/332

Phua, C., Andradi-Brown, D. A., Mangubhai, S., Ahmadia, G. N., Mahajan, S. L., Larsen, K., Friel, S., Reichelt, R., Hockings, M., Gill, D., Veverka, L., Anderson, R., Cédrique Augustave, L., Awaludinnoer Bervoets, T., Brayne, K., Djohani, R., Kawaka, J., Kyne, F., … & Wells, S. (2021). Marine protected and conserved areas in the time of COVID. *PARKS*, *27*, 85–102. https://doi.org/10.2305/IUCN.CH.2021.PARKS-27-SICP.en

Plieninger, T., Hartel, T., Martín-López, B., Beaufoy, G., Bergmeier, E., Kirby, K., Montero, M. J., Moreno, G., Oteros-Rozas, E., & Van Uytvanck, J. (2015). Wood-pastures of Europe: Geographic coverage, social–ecological values, conservation management, and policy implications. *Biological Conservation*, *190*, 70–79. https://doi.org/10.1016/j.biocon.2015.05.014

Reuters News. (2020, July 20). *EU eyes cuts to green transition fund in late bid to strike recovery deal*. https://www.reuters.com/article/uk-eu-summit-climate-change/eu-eyes-cuts-to-green-transition-fund-in-late-bid-to-strike-recovery-deal-idUKKCN24L2IN

Rosa, W. (Ed.). (2017). Transforming our world: The 2030 agenda for sustainable development. In *A new era in global health*. Springer Publishing Company. https://doi.org/10.1891/9780826190123.ap02

SEGOB. (2020, April 23). *DECREE establishing the austerity measures to be observed by the agencies and entities of the Federal Public Administration under the criteria indicated therein (Presidency of the Republic of Mexico)*. https://www.dof.gob.mx/nota_detalle.php?codigo=5592205&fecha=23/04/2020

Simmonds, J. S., Sonter, L. J., Watson, J. E. M., Bennun, L., Costa, H. M., Dutson, G., Edwards, S., Grantham, H., Griffiths, V. F., Jones, J. P. G., Kiesecker, J., Possingham, H. P., Puydarrieux, P., Quétier, F., Rainer, H., Rainey, H., Roe, D., Savy, C. E., Souquet, M., … & Maron, M. (2020). Moving from biodiversity offsets to a target-based approach for ecological compensation. *Conservation Letters*, *13*(2). https://doi.org/10.1111/conl.12695

Singh, R., Galliers, C., Moreto, W., Slade, J., Long, B., Aisha, H., Wright, A., Cartwright, F., Deokar, A., Wyatt, A., Deokar, D., Phoonjampa, R., Smallwood, E., Aziz, R., Benoit, A. K., Cao, R., Willmore, S., Jayantha, D., & Ghosh, S. (2021). Impact of the COVID-19 pandemic on rangers and the role of rangers as a planetary health service. *Parks*, *27*, 119–134.

The Hill News. (2020, November 2). *COVID-19 and the biodiversity crisis*. https://thehill.com/opinion/energy-environment/523944-covid-19-and-the-biodiversity-crisis

UNFCCC. (2016). *The REDD desk*. https://theredddesk.org/what-redd

United Nations. (2020, August 20). *Impact of COVID-19 on SDG progress: A statistical perspective*. https://www.un.org/development/desa/dpad/publication/un-desa-policy-brief-81-impact-of-covid-19-on-sdg-progress-a-statistical-perspective/

Varillas, A. (2020, October 6). *Conanp: Mexico strips its natural protected areas of 75% of their budget*. https://www.eluniversal.com.mx/english/conanp-mexico-strips-its-natural-protected-areas-75-their-budget

Waithaka, J., Dudley, N., Álvarez, M., Mora, S. A., Chapman, S., Figgis, P., Fitzsimons, J., Gallon, S., Gray, T. N. E., Kim, M., Pasha, M. K. S., Perkin, S., Roig-Boixeda, P., Sierra, C., Valverde, A., & Wong, M. (2021). Impacts of COVID-19 on protected and conserved areas: A global overview and regional perspectives. *Parks*, *27*, 41–56.

World Population Review. (2019). *Largest countries in the world 2019*. http://worldpopulationreview.com/countries/largest-countries-in-the-world/

Wunder, S., Brouwer, R., Engel, S., Ezzine-de-Blas, D., Muradian, R., Pascual, U., & Pinto, R. (2018). From principles to practice in paying for nature's services. *Nature Sustainability*, *1*(3), 145–150. https://doi.org/10.1038/s41893-018-0036-x

Xu, H., Cao, Y., Yu, D., Cao, M., He, Y., Gill, M., & Pereira, H. M. (2021). Ensuring effective implementation of the post-2020 global biodiversity targets. *Nature Ecology & Evolution*, *5*, 411–418. https://doi.org/10.1038/s41559-020-01375-y

2　Payments for environmental services

We examine *payments for environmental services* (PES), an important green initiative, for its widespread applications worldwide and well-documented socioecological consequences (Naeem et al., 2015; Wunder, 2005, 2008). This book uses the term *payments for environmental services*, also termed payments for ecosystem services (Wunder, 2015). So we define payments for environmental services as payments made to restore, sustain, or improve ecosystems and the related services that benefit human beings. Payments for environmental services are one kind of essential conservation tool, which have been adopted worldwide to combat global changes that are jeopardizing ecosystem processes, functions, and services at unprecedented rates across the globe (Daily & Matson, 2008).

Under the PES framework, ecosystem service funders pay ecosystem service providers cash or in-kind incentives to change their resource-use behavior, which in turn may restore, maintain, or improve the related ecosystem services that would be unavailable without such payments (Wunder, 2015). For decades, the development and implementation of PES programs sought to combat climate change and environmental degradation (Naeem et al., 2015; Wunder, 2005, 2008). From the PES programs that are "government-financed" (Engel et al., 2008) to those broadly defined by Muradian et al. (2010), both classic and broader definitions are further summarized by Wunder (2015), who reviewed diverse definitions and debates over the past decades.

2.1　PES popularity

PES has witnessed a rapid increase in popularity over the past two decades, with more than 300 PES schemes inventoried worldwide in 2004 (Mayrand & Paquin, 2004) and at least 584 inventoried in 2016 (Ezzine-de-Blas et al., 2016). As of July 2019, the world-renowned PES program Reducing Emissions from Deforestation and Forest Degradation (REDD+) has attracted 39 developing countries to participate, covering a forest area of approximately 1.49 billion ha or 37 percent of the global forest area (Food and Agriculture Organization of the United Nations, 2019). By the end of 2018, the Conservation Reserve Program (CRP), an extensive PES program in the USA, enrolled 9.15 million ha of agricultural land for enhanced ecosystem services (United States Department

DOI: 10.4324/9781003290292-2

of Agriculture & Farm Service Agency, 2019). In their systematic literature search, Ezzine-de-Blas et al. retained a subset of 55 PES programs from hundreds screened worldwide for a global PES tendency assessment (Ezzine-de-Blas et al., 2016). Even this subset covers a total of 615,746 km² (larger than the total area of Ukraine), accounting for an accumulated investment of more than $143.61 billion (Table 1.1). If adding China's National Forest Conservation Program (one of the most extensive PES programs in the world) to this subset, the area rises to 1,785,746 km², larger than the territory of Iran (~1,650,000 km²). The popularity of PES is also evidenced by an exponential increase in PES literature (The Nature Conservancy, 2022), focusing on its conceptual frameworks, principles, design features, participation and compliance, and socio-environmental impacts and trade (Wunder et al., 2018).

Worthy of mention is an article by Wunder, Engel, and Pagiola (2008), which compares and contrasts a number of PES programs and highlights trends among the different methods of operating such projects in terms of, for example, sources of funding, geographic extent and environmental focus, and ways in which PES programs achieve and maintain effective outcomes. Another paper categorizes PES literature between 1974 and 2011, identifying the statistics of PES projects' geographic locations, their economic conceptualization, areas where additional research may benefit PES programs, and PES "types and challenges" in developing and industrialized nations (Schomers & Matzdorf, 2013).

2.2 PES concept

Under the PES concept (Naeem et al., 2015; Wunder et al., 2018), funders, beneficiaries, or representatives make payments to stakeholders to motivate pro-environmental behavior, including reducing resource use and/or pollution. Consequently, the corresponding natural structures, functions, and services are restored, maintained, or enhanced, which would otherwise be impossible (Wunder, 2015). The primary goal of PES is to protect ecosystems and their services of tremendous value to humanity (Friess et al., 2015; Wunder, 2005), including provisioning (e.g., food, water, and fiber), regulating (e.g., climate, floods, and disease), cultural (e.g., recreational, aesthetic, and spiritual benefits), and supporting (e.g., soil formation, photosynthesis, and nutrient cycling) services (Millennium Ecosystem Assessment, 2003). As pointed out earlier, global ecosystem services may have produced an annual monetary value of at least $46 trillion in 1997, and the number increased to $145 trillion in 2011 (both measured in 2007 $US) (Costanza et al., 2014).

2.2.1 Additionality

Additionality is a critical concept in PES literature, which implies that adding a payment to preserve an ecosystem service should benefit a previous baseline in terms of ecosystem services (LaRocco & Deal, 2011). Put another way, paying for an ecosystem service can have additionality if there is a noticeable difference before and after implementation (Gillenwater, 2012). Lack of additionality happens when PES programs pay for an ecosystem service that would have been available regardless of payments, implying a waste of money for no additional ecosystem service (Engel et al., 2008). "Double-dipping" or "piggy-backing" are other terms for lack of additionality.

An excellent example of additionality is the global Reducing Emissions from Deforestation and Forest Degradation (REDD+) program, developed under the Kyoto Protocol to reduce and mitigate greenhouse gas emissions by providing local landowners incentives to protect and promote ecosystem services (UNFCCC, 2016). A common topic in REDD+ is additionality regarding whether implementing a PES contract generates additional benefits or not.

2.2.2 Concurrent PES

Many concurrent PES programs—a prominent type of concurrent green initiatives—bear spillover effects. Concurrent payments for environmental services (concurrent PES hereafter) are widespread in an era when "[t]he biosphere, upon which humanity as a whole depends, is being altered to an unparalleled degree across all spatial scales" (IPBES, 2019). Examples of concurrent PES entail communities that run two or more PES contracts simultaneously in Mexico (Ezzine-de-Blas et al., 2016); households that participate in multiple PES programs in northern Cambodia and China (Clements & Milner-Gulland, 2015; Song et al., 2018); farms funded by multiple concurrent schemes under the same PES umbrella in Costa Rica (LaRocco & Deal, 2011); and farmers' simultaneously signing-up to both the Environmentally Sensitive Area Program and Countryside Stewardship Scheme in the United Kingdom (Hodge & Reader, 2010; Lobley & Potter, 1998). The increasing debate about PES stacking and bundling (see more details in Section 2.3) has also witnessed the popularity of concurrent PES (Cooley & Olander, 2012; Motallebi et al., 2018; World Resources Institute, 2009).

In Australia, the Wet Tropics of Queensland World Heritage Area (WTWHA) was established in 1988 to protect the park area (Harrison et al., 2003). At the same time, the Community Rainforest Reforestation Program (CRRP), established in 1992, was a government-supported, small-area tree-planting program that focuses on private land bordering the WTWHA. The CRRP can complement the WTWHA substantially. Allowing substantial tree farming, the CRRP aimed to offset— likely partially—local social costs that were anticipated when the WTWHA was designated as a protected area. Furthermore, the restoration planting in the CRRP may bring in a co-benefit of enhancing biodiversity in fragmented agricultural landscapes that border WTWHA. When restoring these areas, wildlife corridors and networks—about 60% of CRRP plantings form part of a continuous vegetation corridor network—were established, enhancing the biotic viability in the Arberton Tableland that borders the WTWHA.

In Mexico, the PSA-H (Payment for Ecosystem Services–Hydrological) Program was established in 2003 to avoid the deforestation of parcels in overexploited watersheds (McAfee & Shapiro, 2010). The participants of this program received federal funding, expecting to establish direct contacts between producers and private beneficiaries of hydrological services. The other program, PSA-CABSA (Program for the Development of Markets for the Ecosystem Services of Carbon Sequestration, the Derivatives of Biodiversity, and to Promote the Introduction and Improvement of Agroforestry Systems) was established in 2004. This PSA-CABSA program allowed local communities to develop sustainable management plans, including the production of crops (e.g., coffee, palm, cacao, vanilla, or rubber) that can be "shade grown" in the forest. Payments were made to community associations by the federal government.

Both programs ran concurrently until 2006, when they were consolidated under the larger PROARBOL program. There existed some contradictions between these two programs. For instance, debates existed about whether to promote federal bureaucratic control or to empower rural cooperative associations. The PSA-H program had strong federal control while the PSA-CABSA promoted local community stewardship of the land. In 2004–2005, PSA-H involved around 600 participants protecting over 300,000 hectares of land, while PSA-CABSA had 42 participants protecting almost 60,000 hectares of land. The greater success of PSA-H seemed to be a result of the higher amounts of federal funding. In 2004–2005, PSA-H was allocated approximately US$49 million and PSA-CABSA only an approximate US$13 million.

2.3 Classification of 55 PES programs

To determine whether each of the 55 PES programs in Table 1.1 was/is concurrent with other PES programs and the estimated level of certainty in our decision, we adopted a conservative method. First, we reviewed Ezzine-de-Blas et al. (2016), relevant journal articles, government reports or documents, book chapters, and the like with a keyword of the program name or alternative names. According to our definition, if we found at least one document providing strong evidence that the program under consideration has/had a concurrent program, this program was labeled to have a concurrent program with high certainty. We offer an example to illustrate this evaluation process: if PES programs A and B are explicitly described as being implemented in the same geographical area(s) or having payments made to the same participant(s) simultaneously, we then labeled programs A and B to be concurrent programs with a high level of certainty.

Second, we sought to rely on experts' knowledge for the programs with weak or no evidence of concurrent PES program(s). Individually, we sent email messages to the author(s) of the references we found and/or program managers and asked the following question: "Regarding the PES program in this paper, do you know whether there are other PES or PES-like programs that simultaneously targeted the same site (e.g., land parcel, watershed) and/or enrolled the same participants?" (An et al., in review). If the recipient answered yes or no with certainty, we labeled the program to have or not have a concurrent program(s) with a high level of certainty. If the recipient answered yes or no with some degree of uncertainty, then we labeled a low degree of certainty with a question mark "?".

Worthy of mention is a situation in which no evidence was found regarding a PES program's concurrency with other programs. For instance, none of the contacted experts had such knowledge (e.g., the respondent answered "I don't know"). Under such circumstances, we decided that the program was not concurrent with other programs and therefore labeled the determination with a low level of certainty. The outcome of the above determination is in Table 2.1.

Given the sites identified to have concurrent PES programs in Table 2.1, we sought to provide more information regarding the connection between the concurrent PES programs. Specifically, we reviewed relevant papers we found and provided additional information, including the name of concurrent PES programs and how they are connected. The outcome is Table 2.2. To make Table 2.2 comparable with Table 2.1, we kept all the records (i.e., PES programs) without concurrent PES programs, leaving the two columns "Program 1" and "Program 2" blank and putting the remark "No concurrent payments" in the column "Note".

We found that out of the 55 PES programs identified by Ezzine-de-Blas et al. (2016a), over half of them had concurrent programs (Table 2.2). For all the concurrent PES programs, potential spillover effects vary from site to site (An et al., in review). For instance, the Bolivia case shows that one green initiative refers to payments made by the US Fish and Wildlife Service for bird habitat protection. The concurrent green initiative represents payments for watershed conservation for downstream irrigation, which the local government funds on behalf of downstream irrigators for water stabilization. These two payments were paid to the same farmers in the same watershed (Table 2.2). For other concurrent green initiatives, we refer to Table 2.2. Regardless of differences in details, these green initiatives share the two common features we used to identify concurrent green initiatives: they either cover the same geographic area or make payments to the same recipients.

2.4 PES spillover effects

Spillover effects exist in various PES programs as in generic green initiatives. In this section, we introduce various ways that are used to package (combine) ecosystem services and sell these services as credits (Smith et al., 2015). Along this line, there exists some literature about PES packaging (including bundling and stacking), though the terminology is inconsistent in PES studies. Also, we introduce spillover effects in different regions or countries.

2.4.1 Stacking

Recently, in the USA, a concurrent PES scheme called PES stacking has emerged. However, "there are no regulations addressing stacking or any guidance documents from US federal resource agencies" (Robertson et al., 2014), nor any evidence-based guidelines about how to achieve or improve the intended ecosystem services. Primarily used in North America, stacking (also called layering) refers to providers or landowners receiving multiple payments for multiple ecosystem

Table 2.1 Determination of the 55 PES programs based on information from other sources

Id	Label	Country	Region[a]	Description	E[b]	Con-PES[c]	Certainty	References
1	NegB	Bolivia	S. America	Los Negros, Bolivia		Yes	Certain	Asquith et al. (2008), Robertson and Wunder (2005)
2	PimE	Ecuador	S. America	Pimampiro, Ecuador	√	Yes	Certain	Wunder and Albán (2008)
3	ProE	Ecuador	S. America	PROFAFOR, Ecuador		Yes?	Almost	Wunder and Albán (2008)
4	VitF	France	Europe	Vittel (Nestlé Waters)	√	No	Certain	Perrot-Maître (2005, 2006)
5	SloCh	China	Asia	Sloping Land Conversion		Yes	Certain	Bennett (2008), Xu al. (2010)
6	HydroMX	Mexico	N. America	Payment for Ecosystem Services–Hydrological (PSA-H), Mexico	√	Yes	Certain	Muñoz-Piña et al. (2008). Alix-Garcia et al. (2012), Corbera et al. (2009)
7	ConsUS	USA	N. America	Conservation Reserve Program (CRP), USA	√	Yes	Certain	Claassen et al. (2008), Leathers and Harrington (2000)
8	EnvUS	USA	N. America	Environmental Quality Incentives Program (EQIP)	√	Yes	Certain	Claassen et al. (2008), Cattaneo (2003)
9	SenUK	United Kingdom	Europe	Environmentally Sensitive Area (ESA)	√	No	Certain	Dobbs and Pretty (2008), Crabtree et al. (2000)
10	CouUK	United Kingdom	Europe	Countryside Stewardship Scheme	√	No	Certain	Dobbs and Pretty (2008), Crabb et al. (2000)

(Continued)

Table 2.1 Continued

Id	Label	Country	Region[a]	Description	E[b]	Con-PES[c]	Certainty	References
11	NotrG	Germany	Europe	Northeim model project	√	No?	Almost	Klimeka et al. (2008), Grolleau and McCann (2012)
12	CamZ	Zimbabwe	Africa	CAMPFIRE, Zimbabwe		No?	Almost	Frost and Bond (2008), Dunham et al. (2003), Harrison (2015)
13	LavCo	Colombia	S. America	CIPAV-Río La Vieja		Yes?	Almost	Pagiola et al. (2010)
14	ChaE	Ecuador	S. America	Chachis, Ecuador	√	Yes	Certain	Wendland and Suárez (2009), Speiser et al. (2009)
15	ChaCo	Colombia	S. America	Chaina, Colombia		No?	Unsure	Moreno-Sanchez et al. (2012), Dillaha et al. (2007)
16	ProcCo	Colombia	S. America	Procuenca, Colombia	√	No?	Unsure	Dillaha et al. (2007), Erazo and Benjumea (2004)
17	CelE	Ecuador	S. America	Celica, Ecuador	√	No	Certain	Raes et al. (2012)
18	ChacE	Ecuador	S. America	Chaco, Ecuador		Yes?	Almost	Cordero-Camacho (2008)
19	RichS	Sudáfrica	S. Africa	Richtersveld		No?	Unsure	Robinson (1998)
20	KitK	Kenya	E. Africa	Kitengela, Kenya	√	Yes?	Unsure	Yatich et al. (2008), Osano et al. (2012)
21	MenM	Madagascar	E. Africa	Menabe, Madagascar	√	No	Certain	Sommerville et al. (2010)
22	TurtT	Tanzania	E. Africa	Sea turtle nest, Tanzania		No?	Almost	Ferraro (2007), Ferraro and Gjertsen (2009)
23	BirwCa	Cambodia	Asia	Bird watch & ecotourism		Yes	Certain	Clements et al. (2010), Clements et al. (2008)
24	BirnCa	Cambodia	Asia	Bird nest protection		Yes	Certain	Clements et al. (2010), Clements et al. (2013)
25	SocE	Ecuador	S. America	Socio Bosque	√	Yes	Certain	de Koning et al. (2011), Krause et al. (2013), Holland et al. (2014)

26	JesH	Honduras	C. America	Jesús de Otoro, Honduras		No?	Almost	Kosoy et al. (2007)
27	HerCR	Costa Rica	C. America	Heredia, Costa Rica		Yes	Certain	Kosoy et al. (2007), Pagiola (2008)
28	SanNC	Nicaragua	C. America	San Pedro del Norte		No?	Almost	Kosoy et al. (2007)
29	KulN	Nepal	S. Asia	Kulekhani, Nepal	√	No	Certain	Khatri (2009), Joshi (2011)
30	DonV	Vietnam	Asia	Da Nhim PWS, Dong Nai watershed	√	Yes	Certain	To et al. (2012)
31	SonV	Vietnam	Asia	Son La PWS	√	Yes	Certain	To et al. (2012)
32	OacI	India	S. Asia	Oach Kalan-Kuhan mini micro watershed		No?	Unsure	Agarwal et al. (2007)
33	ZapMX	Mexico	N. America	Saltillo, Zapaliname		No?	Almost	Wunder and Wertz-Kanounnikoff (2009)
34	SimT	Tanzania	East Africa	Simanjiro valley	√	No	Certain	Nelson (2008)
35	BioN	Namibia	S. Africa	NRCB management	√	No	Certain	Weaver and Petersen (2008), Naidoo et al. (2016)
36	CatUS	USA	N. America	Catskills, NYC, USA	√	No?	Almost	Grolleau and McCann (2012)
37	UlgT	Tanzania	E. Africa	PWS, Tanzania	√	Yes	Certain	Lopa et al. (2012), Branca et al. (2011)
38	SilNC	Nicaragua	C. America	Silvopastoril, Nicaragua		Yes?	Almost	Pagiola et al. (2008), Pagiola et al. (2017)
39	SilCR	Costa Rica	C. America	Silvopastoril		Yes	Certain	Ibrahim et al. (2007), Pagiola (2010)
40	CidIS	Indonesia	S. Asia	Cidanau watershed PES scheme		No?	Almost	Leimona et al. (2010), Suyanto et al. (2005)
41	TreU	Uganda	E. Africa	Uganda, Trees for Global Benefits Programme	√	No	Certain	German et al. (2010), Carter (2009), Fisher et al. (2018)
42	FirPH	Philippines	A.	Philippines, No Fire Bonus Scheme	√	No	Certain	Soriaga and Annawi (2010)

(*Continued*)

Table 2.1 Continued

Id	Label	Country	Region[a]	E[b]	Description	Con-PES[c]	Certainty	References
43	TreMZ	Mozambique	E. Africa	√	Mozambique, carbono.	Yes?	Almost	Hegde (2010), Jindal (2012), Hegde et al. (2015)
44	TreMX	Mexico	N. America	√	Scolel Té	Yes	Certain	Hendrickson and Corbera (2015), Tipper (2002)
45	MonMX	Mexico	N. America	√	Monarch	Yes	Certain	Honey-Rosés et al. (2011), Honey-Rosés et al. (2009)
46	PwsG	Germany	Europe	√	PWS en Munich	No?	Almost	Grolleau and McCann (2012)
47	SumIn	Indonesia	S. Asia		Sumberjaya AF conservation auction	Yes?	Almost	Leimona et al. (2009), Suyanto (2007)
48	RefK	Kenya	E. Africa		Reforestation conservation auction	No?	Unsure	Khalumba et al. (2014)
49	FuqCo	Colombia	S. America		Fuquene, Colombia	Yes?	Almost	Quintero and Otero (2006)
50	KmpfB	Bolivia	S. America		Noel Kempff Mercado REDD+ project	Yes?	Almost	Asquith et al. (2002), Pereira (2010), Grieg-Gran et al. (2005)
51	PesCR	Costa Rica	C. America		Payments for Environmental Services (PSA)	Yes	Certain	Pagiola (2008)
52	MunCo	Colombia	S. America		CIPAV- La Salvajina & PNN Munchique	No?	Unsure	CIPAV (2007)
53	MakMa	Madagascar	E. Africa		Makira WCS	No	Certain	Brimont et al. (2015)
54	LrcB	Bolivia & Peru	S. America	√	Landrace conservation payments	Yes?	Almost	Narloch et al. (2011)
55	BfeB	Brazil	S. America	√	Bolsa Floresta	No	Certain	Börner et al. (2013)

Notes:
[a] Under the column for Region, E stands for East, W for West, S for South, N for North, and C for Central.
[b] The column labeled with E stands for whether we contacted expert(s)—often the author(s) of the reference(s)—and received responses.
[c] Con-PES refers to whether there is a concurrent PES program existing with the PES program.

Table 2.2 Description of concurrent payments for the 55 PES programs

Id	Label	Country	Program 1	Program 2	Note
1	NegB	Bolivia	Payment for bird habitat protection	Payment for watershed conservation for downstream irrigation	Two buyers paid the same farmers in the same watershed: one is the US Fish and Wildlife Service to protect bird habitat; the other is the local government on behalf of downstream irrigators for water stabilization
2	PimE	Ecuador	Pimampiro (watershed protection)	PROFAFOR (carbon sequestration)	The same watershed was protected for both water stabilization (Policy 1) and carbon sequestration (Policy 2) through forest protection and afforestation
3	ProE	Ecuador	PROFAFOR (carbon sequestration)	Pimampiro (watershed protection)	The same watershed was protected for both carbon sequestration (Policy 1) and water stabilization (Policy 2) through forest protection and afforestation
4	VitF	France			No concurrent payments
5	SloCh	China	SLCP or GTGP (water and soil conservation)	NFCP or EWFP (water and soil conservation)	The central government paid the same households to plant trees on cropland with steeping slopes (Policy 1) and conserve natural forests (Policy 2)
6	HydroMX	Mexico	PSA-H (hydrological services)	SA-CABSA (carbon, biodiversity, and agro-forestry services)	The same recipients in the same watershed: the two PES programs were later merged into a single policy framework
7	ConsUS	USA	CRP (soil conservation)	EQIP (multiple objects beyond soil conservation)	Same recipients: an individual farmer could have some land enrolled in the CRP and other lands (or livestock) enrolled in EQIP, i.e., having contracts under more than one program

(*Continued*)

Table 2.2 Continued

Id	Label	Country	Program 1	Program 2	Note
8	EnvUS	USA	EQIP (multiple objects beyond soil conservation)	CRP (soil conservation)	Same recipients: an individual farmer could have some land enrolled in the CRP and other land (or livestock) enrolled in EQIP, i.e., having contracts under more than one program
9	SenUK	United Kingdom			No concurrent payments
10	CouUK	United Kingdom			No concurrent payments
11	NotrG	Germany			No concurrent payments
12	CamZ	Zimbabwe			No concurrent payments
13	LavCo	Colombia	Silvopastoral practice-1 (biodiversity conservation and carbon sequestration)	Silvopastoral practice-2 (biodiversity conservation and carbon sequestration)	Likely same households would receive multiple payments if adopting several practices such as planting trees and/or shrubs for feeding livestock and fencing as wind screens
14	ChaE	Ecuador	Payment in Gran Reserva Chachi (biodiversity conservation)	Socio Bosque (carbon storage, biodiversity protection, water provision)	Same geographic region by the local rewards for biodiversity (Policy 1) and national PES program (Policy 2)
15	ChaCo	Colombia			No concurrent payments
16	ProcCo	Colombia			No concurrent payments
17	CelE	Ecuador			No concurrent payments
18	ChacE	Ecuador	Payment for watershed conservation	Socio Bosque (carbon storage, biodiversity protection, water provision)	Possibly same watershed for water conservation (Policy 1) and other services by national program (Policy 2)
19	RichS	Sudáfrica			No concurrent payments

			WLP (wildlife conservation)	OOC (Olare Orok Conservancy)	
20	KitK	Kenya	WLP (wildlife conservation)	OOC (Olare Orok Conservancy)	Likely same geographic region: WLP for wildlife conservation later continued and expanded to cover Maasai Mara National Reserve where OOC was implemented
21	MenM	Madagascar			No concurrent payments
22	TurtT	Tanzania			No concurrent payments
23	BirwCa	Cambodia	Payment for community-based Ecotourism	Agri-environment payments and payment for bird nest protection	Same recipients in the same village by three PES programs for bird nest protection, agri-environmental conservation, and ecotourism
24	BirmCa	Cambodia	Payment for bird nest protection	Agri-environment payments and payment for community-based Ecotourism	Same recipients in the same village by three PES programs for bird nest protection, agri-environmental conservation, and ecotourism
25	SocE	Ecuador	Socio Bosque (carbon storage, biodiversity protection, water provision)	E.g., Payment in Gran Reserva Chachi (biodiversity conservation)	Socio Bosque is a national PES program, covering many geographic regions and compensating individuals participating in other PES programs
26	JesH	Honduras			No concurrent payments
27	HerCR	Costa Rica	Payment-1 for watershed conservation	Payment-2 for watershed conservation	Same watershed by two users with stacking payments: one is downstream water user; the other is a beverage company (Florida Ice and Farm Co.)
28	SanNC	Nicaragua			No concurrent payments
29	KulN	Nepal			No concurrent payments
30	DonV	Vietnam	Payment for watershed conservation	Payment for Forest Protection and Planting	There exist multiple programs in the same geographic region in Vietnamese uplands, particularly in the north-west mountainous area

(Continued)

Table 2.2 Continued

Id	Label	Country	Program 1	Program 2	Note
31	SonV	Vietnam	Payment for watershed conservation	Payment for Forest Protection and Planting	There exist multiple programs in the same geographic region in Vietnamese uplands, particularly in the north-west mountainous area
32	OacI	India			No concurrent payments
33	ZapMX	Mexico			No concurrent payments
34	SimT	Tanzania			No concurrent payments
35	BioN	Namibia			No concurrent payments
36	CatUS	USA			No concurrent payments
37	UlgT	Tanzania	PWS (watershed)	Payment for bird protection	Same watershed in Uluguru Mountains by two payments: one operated by CARE & WWF for downstream water; the other funded by UK DFID and run by the UK royal society for bird protection
38	SilNC	Nicaragua	Silvopastoral practice-1 (biodiversity conservation and carbon sequestration)	Silvopastoral practice-2 (biodiversity conservation and carbon sequestration)	Likely same households would receive multiple payments if adopting several practices such as planting trees and/or shrubs for feeding livestock and fencing as wind screens
39	SilCR	Costa Rica	Silvopastoral practice-1 (biodiversity conservation and carbon sequestration)	Silvopastoral practice-2 (biodiversity conservation and carbon sequestration)	The same households would receive multiple payments if adopting several practices such as planting trees and/or shrubs for feeding livestock and fencing as wind screens
40	CidIS	Indonesia			No concurrent payments
41	TreU	Uganda			No concurrent payments
42	FirPH	Philippines			No concurrent payments

No.	Code	Country	Payment for carbon	Payment for biodiversity	Notes
43	TreMZ	Mozambique	Payment for carbon	Payment for biodiversity	Bundled payment model, which later might evolve to multiple payments on same geographic region or to same recipients
44	TreMX	Mexico	Payment for hydrological services (water)	Payment for forestry services (carbon sequestration)	Same participants for hydrological services (Payment-1) and forestry services (Payment-2)
45	MonMX	Mexico	MBCF (Monarch butterfly conservation Fund)	Payments for conservations ("fondos concurrentes")	In the same region of Monarch Butterfly Biosphere Reserve, other PES programs with different fund sources were operated, colloquially referred to as "fondos concurrentes"
46	PwsG	Germany			No concurrent payments
47	SumIn	Indonesia	Conditional tenures (secure tenure with forestry permits)	RiverCare (watershed functions of reducing sediment)	Likely the same watershed region under two projects: in the first project, government use secure tenure as in-kind payment for forest protection from deforestation; the second for dam functions by reducing sediment
48	RefK	Kenya			No concurrent payments
49	FuqCo	Colombia	CPWF fund for solid and water conservation	GTZ's CAP (sustainable land use for watershed)	The same watershed region with two projects
50	KmpfB	Bolivia	NKM Climate Action Project (carbon mitigation)	Payment for "bundled" biodiversity along with carbon	"Bundled" initiative combining carbon and biodiversity benefits by US Initiative on Joint Implementation and The Nature Conservancy and a consortium of US companies
51	PesCR	Costa Rica	PSA (mainly water conservation)	E.g., GEF (payments from biodiversity users)	Likely same recipients within same watersheds with multiple fund sources including fossil fuel sales tax revenues, and biodiversity users; stacking model
52	MunCo	Colombia			No concurrent payments

(*Continued*)

Table 2.2 Continued

Id	Label	Country	Program 1	Program 2	Note
53	MakMa	Madagascar			No concurrent payments
54	LrcB	Bolivia & Peru	PACS (agrobiodiversity conservation)	Other multiple incentives for generic diversity	Probably same site in Puro (Peru) with multiple incentives to support sustainable agricultural production for genetic diversity conservation
55	BfeB	Brazil			No concurrent payments

Note: The references are the same as those in Table 2.1.

services they supply. Unlike bundling, each ecosystem service is sold separately rather than together.

One PES program can offer both bundling and stacking options, but they cannot be performed in unison on one site because stacking requires that services become unbundled (Gillenwater, 2012). Combining payments and ecosystem services is beneficial through stacking and may be better for sellers mainly because they receive the highest amount of payments and are likely to produce the most substantial amount of ecosystem services (Gillenwater, 2012; Hejnowicz et al., 2014). Stacking payments for environmental services may also allow landowners to take on more extensive projects that would have otherwise not been economically feasible (Gillenwater, 2012; Hejnowicz et al., 2014). Another potential benefit of stacking is diversifying buyers (Hejnowicz et al., 2014). Research shows that if a landowner receives payment for only one of the ecosystem services, s/he may feel it is a financially unachievable project, but if payments could be stacked, the project might be implementable (Gillenwater, 2012).

Ecosystem services from one land parcel can be stacked and sold to more buyers. However, most literature—especially in the USA—focuses on one buyer purchasing stacked ecosystem services (often from one seller). There are three primary forms of stacking: horizontal, vertical, and temporal (Cooley & Olander, 2012). Horizontal stacking implies that when landowners participate in multiple conservation projects on different land areas, they receive a payment for each ecosystem service derived from each area of land. Vertical stacking implies when a landowner does one conservation project on one land area and receives multiple payments for the multiple ecosystem services derived from that area (Cooley & Olander, 2012). Temporal stacking is like vertical stacking, where a landowner implements only one conservation program but receives payments for different ecosystem services over time as payment programs develop. Horizontal stacking is uncontroversial, but vertical and temporal stacking has ignited debate due to their potential for "double counting" (also called "double-dipping" or "piggybacking") (Gillenwater, 2012; Hejnowicz et al., 2014; Smith et al., 2015). These terms are also more formally known as a lack of additionality.

2.4.2 Bundling

"Bundling" occurs when multiple ecosystem services generated within a land parcel are sold together as one commodity to a buyer (usually from one seller). In other words, sellers earn one payment for multiple ecosystem services (Cooley & Olander, 2012). The potential benefits of bundling are that it allows providers to receive payments for multiple ecosystem services generated as byproducts of an overarching ecosystem service. For example, by managing land to improve forest habitat for wildlife, many other ecosystem services can be created, such as carbon sequestration, scenic beauty, increased soil integrity, and water filtration.

Bundling "may be the best way of securing a sale and avoiding free-riding" (Smith et al., 2015). Bundling is desirable when the conservation goals of a PES program are broad (Hejnowicz et al., 2014). Furthermore, bundling can potentially reduce organizational costs and increase payouts to participants (Hejnowicz et al., 2014). While proponents of bundling claim it "recognizes the interconnectedness of ecosystem services", opponents state that it may be too difficult to measure and manage multiple ecosystem services at once (Hejnowicz et al., 2014).

2.4.3 Policy implementation

PES policies—or green initiatives in a broader sense—must be designed and implemented at multiple local jurisdiction levels. However, there may exist some level of inconsistencies in this regard. Take Australia as an example. There is a federal system of urban governance in which planning legislation and policy framework are set by six states and two territories and implemented by more than 500 local governments through their land use plans. These local plans are prepared under state planning legislation. However, due to varying nomenclature, local instruments combine a mix of policy objectives with concrete provisions for spatial allocation of land uses (Gurran et al., 2015).

Australian national urban policy articulates high-level principles for settlement planning by states and territories that emphasize the need for mixed urban centers, biodiversity conservation, and sustainable design. However, there is no guarantee that state mandates for sustainability will result in local implementation through planning instruments or decisions.

An analysis was performed that focused on the plans prepared up to 2009, some of which were reaching completion in 2013. The results demonstrated considerable heterogeneity in local planning schemes, despite ongoing planning system reforms across Australia which sought to standardize local plans. Metropolitan local government areas displayed spatial differences in sustainability policy adoption (Gurran et al., 2015).

This is a fairly extensive study of the tools and conservation measures, planning approaches, land uses, and policy implementations that are used in very different ways at the local level. The main idea, as far as policy interactions, is that an umbrella policy can be implemented at the state or federal level, but can be applied very differently in different local jurisdictions.

2.5 Grand challenges

Many challenges—be they loss and fragmentation of forest areas, biodiversity loss, wildlife extinction, desertification, and the like—are jeopardizing humanity at unprecedented rates from local to global scales. Thus, they can be called grand challenges. Virtually all these grand challenges can be traced back to various human activities in the context of increasing population pressure. Humans are degrading or destroying ecosystems rapidly, threatening the very "life-support services of tremendous value" such as food, water, clean air, soil, and forests (Daily & Matson, 2008). Many protected areas—such as national parks and nature reserves—are not exempted from such degradation (Curran et al., 2004;

Liu et al., 2001). To address such challenges, the International Convention of Biological Diversity's Aichi targets (https://www.cbd.int/sp/target/) have called for protecting natural habitats (Target 5), threatened species (Target 12), and various ecosystem services from natural ecosystems (Target 14). The United Nations' 17 Sustainable Development Goals, especially Goal 15, aim to protect, restore, and promote sustainable use of terrestrial ecosystems (United Nations, 2016).

In this context, payments for ecosystem services (PES) have come into being for decades, aiming to provide incentives directly to resource users to take environmentally beneficial actions or to refrain from environmentally harmful actions in the hope of protecting ecosystems and the related services. Although PES programs have been reported to restore ecosystems and improve human well-being successfully, many challenges have surfaced in many PES programs.

First, PES programs suffer from lacking sustainability. Many participants return to their pre-PES behavioral patterns once PES payments become no longer available. This problem is widely observable globally, including both developing (Uchida et al., 2005) and developed countries (Claassen et al., 2008). Current PES research pays attention to individual factors such as farm income, land quality, land plot slope, distance from household to the land parcel, age of contract holders, labor supply, and livelihood alternatives (Adhikari & Agrawal, 2013; Engel et al., 2008; He & Sikor, 2015). These variables are primarily treated in a piecemeal manner, while the feedback loops and nonlinear relationships are largely overlooked. Also importantly, there is a dire need to measure the environmental performance of PES programs. So far, the most used measure is land use and land cover (LULC), and very few PES programs have paid enough attention to faunal and/or floral changes in responses to PES programs—there are several exceptions (e.g., Liu et al., 2008; Tuanmu et al., 2016). Therefore, PES research and implementation must consider "the complex interrelationship among socioeconomic, demography and ecological metrics" on the one hand while developing and testing more representative ecological metrics on the other hand (Lewison et al., 2017).

A critical line in PES research is to compare PES to other conservation tools such as protected areas and community-based natural resource management (Börner et al., 2017). For instance, Robalino and collaborators found that a PES program had little additionality in and around protected areas in Costa Rica (Robalino et al., 2015). A case study in Mexico found that the benefits of PES relied heavily on community training and involvement during the implementation of community forest management (Börner et al., 2017).

References

Adhikari, B., & Agrawal, A. (2013). Understanding the social and ecological outcomes of PES projects: A review and an analysis. *Conservation and Society, 11*(4), 359–374.

Agarwal, C., Tiwari, S., Borgoyary, M., Acharya, A., & Morrison, E. (2007). *Fair deals for watershed services in India. Natural resource issues no. 10*. International Institute for Environment and Development.

Alix-Garcia, J. M., Shapiro, E. N., & Sims, K. (2012). Forest conservation and slippage: Evidence from Mexico's National payments for ecosystem services program. *Land Economics*, *88*(4), 613–638.

An, L., Liu, J., Zhang, Q., Song, C., Ezzine-de-Blas, D., Dai, J., Zhang, H., Lewison, R., Bohnett, E., Stow, D., Xu, W., & Bryan, B. A. (in review). Global hidden spillover effects among concurrent green initiatives. *Biological Conservation*.

Asquith, N. M., Vargas, M. T., & Wunder, S. (2008). Selling two environmental services: In-kind payments for bird habitat and watershed protection in Los Negros, Bolivia. *Ecological Economics*, *65*, 675–684.

Asquith, N. M., Vargas-Ríos, M. T., & Smith, J. (2002). Can forest-protection carbon projects improve rural livelihoods? Analysis of the Noel Kempff Mercado climate action project, Bolivia. *Mitigation and Adaptation Strategies for Global Change*, *7*, 323–337.

Bennett, M. T. (2008). China's sloping land conversion program: Institutional innovation or business as usual? *Ecological Economics*, *65*, 699–711.

Börner, J., Baylis, K., Corbera, E., Ezzine-de-Blas, D., Honey-Rosés, J., Persson, U. M., & Wunder, S. (2017). The Effectiveness of payments for environmental services. *World Development*, *96*, 359–374. https://doi.org/10.1016/j.worlddev.2017.03.020

Börner, J., Wunder, S., Reimer, F, Bakkegaard, R. K., Viana, V., Tezza, J., Pinto, T., Lima, L., & Marostica, S. (2013). *Promoting forest Stewardship in the Bolsa Floresta Programme: Local livelihood strategies and preliminary impacts*. Center for International Forestry Research (CIFOR), Fundação Amazonas Sustentável(FAS). Zentrum für Entwicklungsforschung (ZEF), University of Bonn.

Branca, G., Lipper, L., Neves, B., Lopa, D., & Mwanyoka, I. (2011). Payments for watershed services supporting sustainable agricultural development in Tanzania. *The Journal of Environment & Development*, *20*(3), 278–302.

Brimont, L., Ezzine-de-Blas, D., Karsenty, A., Tambaza, F., Toulon, A., Rasolofonirina, G., Razanamihanta, E. (2015). Achieving REDD+ objectives along with equity principles in Madagascar: a critical analysis of Makira project's development strategy. *Forests*, *6*, 748–768.

Carter, S. (2009). *Socio-economic benefits in Plan Vivo projects: trees for global benefits, Uganda*. Plan Vivo Foundation and ECOTRUST.

Cattaneo, A. (2003). The pursuit of efficiency and its unintended consequences: contract, withdrawals in the environmental quality incentives program. *Review of Agricultural Economics*, *25*, 449–469.

CIPAV. (2007). *Desarrollo del pago por servicios ambientales para la conservacion y restauración de ecosistemas en el corredor biológico y multicultural Munchique Pinche*. CIPAV.

Claassen, R., Cattaneo, A., & Johansson, R. (2008). Cost-effective design of agri-environmental payment programs: U.S. experience in theory and practice. *Ecological Economics*, *65*(4), 737–752.

Clements, T., & Milner-Gulland, E. J. (2015). Impact of payments for environmental services and protected areas on local livelihoods and forest conservation in northern Cambodia. *Conservation Biology*, *29*(1), 78–87. https://doi.org/10.1111/cobi.12423

Clements, T., Ashish, J., Nielsen, K., An, D., Tan, S., & Milner-Gulland, E. J. (2010). Payments for biodiversity conservation in the context of weak institutions: Comparison of three programs from Cambodia. *Ecological Economics*, *69*, 1283–1291.

Clements, T., John, A., Nielsen, K., Vicheka, C., Sokha, E., & Piseth, M. (2008). *Tmatboey community-based ecotourism project, Cambodia* (p. 56). USAID, TransLinks.

Clements, T., Rainey, H., An, D., Rours, V., Tan, S., Thong, S., Sutherland, W. J., Milner-Gulland, E. J. (2013). An evaluation of the effectiveness of a direct payment for biodiversity conservation: The Bird Nest Protection Program in the Northern Plains of Cambodia. *Biological Conservation, 157*, 50–59.

Cooley, D., & Olander, L. (2012). Stacking ecosystem services payments: Risks and solutions. *Environmental Law Reporter, 42*, 10150–10165.

Corbera, E., Soberanis, C. G., & Brown, K. (2009). Institutional dimensions of payments for ecosystem services: An analysis of Mexico's carbon forestry programme. *Ecological economics, 68*(3), 743–761.

Cordero-Camacho, D. (2008). Esquemas de pagos por servicios ambientales para la conservación de cuencas hidrográficas en el Ecuador. *Investigación Agraria: Sistemas y Recursos Forestales, 17*(1), 54–66.

Costanza, R., de Groot, R., Sutton, P., van der Ploeg, S., Anderson, S. J., Kubiszewski, I., Farber, S., & Turner, R. K. (2014). Changes in the global value of ecosystem services. *Global Environmental Change, 26*, 152–158. https://doi.org/10.1016/j.gloenvcha.2014.04.002

Crabb, J., Short, C., Temple, M., Winter, M., Augustin, B., & Dauven, A. (2000). Economic evaluation of the countryside stewardship scheme. Prepared for Ministry of Agriculture, Fisheries and Food (MAFF), Economics (Resource Use) Division. Europe, UK: Cheltenham and Glouchester College of Higher Education and ADAS.

Crabtree, B., Thorburn, A., Chalmers, N, Roberts, D, Wynn, G, Barron, N, & Macmillan, D, Barraclough, F. (2000). *Socio-economic and agricultural impacts of the environmentally sensitive areas (ESA) scheme in Scotland. A report for the Scottish executive rural affairs department. Macaulay land use research Institute, with Bell-Ingram rural and university of Aberdeen, Craigiebuckler, Aberdeen.*

Curran, L. M., Trigg, S. N., McDonald, A. K., Astiani, D., Hardiono, Y. M., Siregar, P., Caniago, I., & Kasischke, E. (2004). Lowland forest loss in protected areas of Indonesian Borneo. *Science, 303*(5660), 1000–1003.

Daily, G. C., & Matson, P. A. (2008). Ecosystem services: From theory to implementation. *Proceedings of the National Academy of Sciences of the United States of America, PNAS, 105*(28), 9455–9456. https://doi.org/10.1073/pnas.0804960105

de Koning, F., Aguiñaga, M., Bravo, M., Chiu, M., Lascano, M., Lozada, T., & Suarez, L. (2011). Bridging the gap between forest conservation and poverty alleviation: the Ecuadorian Socio Bosque program. *Environmental Science & Policy, 14*, 531–542.

Dillaha, T. A., Ferraro, P. J., Huang, M., Southgate, D., Upadhyaya, S. K., & Wunder, S. (2007). *Payments for watershed services: Regional syntheses.* Washington, DC, USA: USAID PES Brief 7

Dobbs, T. L., & Pretty, J. (2008). Case study of agri-environmental payments: The United Kingdom. *Ecological Economics, 65*, 765–775.

Dunham, K. M., Davies, C., & Muhwandagara, K. (2003). *Area and quality of wildlife habitat in selected CAMPFIRE districts.* WWF-SARPO and the CAMPFIRE Association, Harare, Zimbabwe (mimeo).

Engel, S., Pagiola, S., & Wunder, S. (2008). Designing payments for environmental services in theory and practice: An overview of the issues. *Ecological Economics, 65*, 663–674.

Erazo, J., & Benjumea, F. (2004). *Análisis de la aplicación de la exoneración del impuesto predial como incentivo para la conservación en Manizales* (p. 38). Instituto Von Humbodt.

Ezzine-de-Blas, D., Wunder, S., Ruiz-Pére, M., & Moreno-Sanchez, R. del P. (2016). Global patterns in the implementation of payments for environmental services. *PLOS ONE, 11*(3), e0149847https://doi.org/10.1371/journal.pone.0149847

Ferraro, P. J. (2007). Performance payments for sea turtle nest protection in low-income nations: A case study from Tanzania. Southwest Fisheries Science Center, National Marine Fisheries Service National Oceanic and Atmospheric Administration. Available at http://www2.gsu.edu/~wwwcec/docs/doc%20updates/NOAA%20Paper%20TZ%20Final%20Draft%20June%202007.pdf

Ferraro, P. J., & Gjertsen, H. (2009). A global review of incentive payments for sea turtle conservation. *Chelonian Conservation and Biology*, *8*(1), 48–56.

Fisher, J. A., Cavanagh, C. J., Sikor, T., & Mwayafu, D. M. (2018). Linking notions of justice and project outcomes in carbon offset forestry projects: Insights from a comparative study in Uganda. *Land Use Policy*, *73*, 259–268.

Food and Agriculture Organization of the United Nations. (2019). *From reference levels to results reporting: REDD+ under the United Nations Framework Convention on Climate Change (no. 9)*. Food and Agriculture Organization of the United Nations.

Friess, D. A., Phelps, J., Garmendia, E., & Gómez-Baggethun, E. (2015). Payments for Ecosystem Services (PES) in the face of external biophysical stressors. *Global Environmental Change*, *30*, 31–42. https://doi.org/10.1016/j.gloenvcha.2014.10.013

Frost, P. G. H., & Bond, I. (2008). The CAMPFIRE programme in Zimbabwe: Payments for wildlife services. *Ecological Economics*, *65*, 776–787.

German, L. A., Ruhweza, A., Mwesigwa, R., Kalanzi, C. (2010). Social and environmental footprints of carbon payments: A case study from Uganda. In H. Suich, L. Tacconi, & S. Mahanty (Eds.), *Payments for environmental services, forest conservation and climate change livelihoods in the REDD?*. Edward Elgar

Gillenwater, M. (2012). *What is additionality? Part 3: Implications for stacking and unbundling*. Greenhouse Gas Management Institute.

Grieg-Gran, M., Porras, I., & Wunder, S. (2005). How can market mechanisms for forest environmental services help the poor? Preliminary lessons from Latin America. *World Development*, *33*(9), 1511–1527.

Grolleau, G., McCann, L. M. J. (2012). Designing watershed programs to pay farmers for water quality services: Case studies of Munich and New York City. *Ecological Economics*, *76*, 87–94.

Gurran, N., Gilbert, C., & Phibbs, P. (2015). Sustainable development control? Zoning and land use regulations for urban form, biodiversity conservation and green design in Australia. *Journal of Environmental Planning and Management*, *58*(11), 1877–1902. https://doi.org/10.1080/09640568.2014.967386

Harrison, E. P. (2015). Impacts of natural resource management programmes on rural livelihoods in Zimbabwe–the ongoing legacies of CAMPFIRE. In PSA Conference. Zimbabwe, Africa

Harrison, R., Wardell-Johnson, G., & Mcalpine, C. (2003). Rainforest reforestation and biodiversity benefits: A case study from the Australian wet tropics. *Annals of Tropical Research*, *25*(2), 65–75.

He, J., & Sikor, T. (2015). Notions of justice in payments for ecosystem services: Insights from China's Sloping Land Conversion Program in Yunnan Province. *Land Use Policy*, *43*, 207–216. https://doi.org/10.1016/j.landusepol.2014.11.011

Hegde, R. (2010). *Payments for ecosystem services and farm household behaviour: The case of carbon in Mozambique's Agroforests*. PhD University of British Columbia.

Hegde, R., Bull, G. Q., Wunder, S., & Kozak, R. A. (2015). Household participation in a payments for environmental services programme: The Nhambita Forest Carbon Project (Mozambique). *Environment and Development Economics*, *20*(5), 611–629.

Hejnowicz, A. P., Raffaelli, D. G., Rudd, M. A., & White, P. C. L. (2014). Evaluating the outcomes of payments for ecosystem services programmes using a capital asset framework. *Ecosystem Services*, *9*, 83–97.

Hendrickson, C. Y., & Corbera, E. (2015). Participation dynamics and institutional change in the Scolel Té carbon forestry project, Chiapas, Mexico. *Geoforum*, *59*, 63–72.

Hodge, I., & Reader, M. (2010). The introduction of entry level Stewardship in England: Extension or dilution in agri-environment policy? *Forest Transitions*, *27*(2), 270–282. https://doi.org/10.1016/j.landusepol.2009.03.005

Holland, M. B., De Koning, F., Morales, M., Naughton-Treves, L., Robinson, B. E., & Suárez, L. (2014). Complex tenure and deforestation: implications for conservation incentives in the Ecuadorian Amazon. *World Development*, *55*, 21–36.

Honey-Rosés, J., Baylis, K., & Ramirez, I. (2011). A spatially explicit estimate of avoided forest loss. *Conservation Biology*, *25*(5), 1032–1043.

Honey-Rosés, J., Lopez-Garcia, J., Rendon-Salinas, E., Peralta-Higuera, A., & Galindo-Leal, C. (2009). To pay or not to pay? Monitoring performance and enforcing conditionality when paying for forest conservation in Mexico. *Environmental Conservation*, *36*(2), 120–128.

Ibrahim, M., Gobbi, J., Casasola, F., Chacón, M., Ríos, N., Tobar, D., Villanueva, C., & Sepúlveda, C. (2007). *Enfoques alternativos de pagos por servicios ambientales: Experiencia del proyecto Silvopastoril.* World Bank.

IPBES. (2019). *Summary for policymakers of the global assessment report on biodiversity and ecosystem services of the intergovernmental science-policy platform on biodiversity and ecosystem services.* Intergovernmental Science-Policy Platform on Biodiversity and Ecosystem Services. https://www.ipbes.net/news/ipbes-global-assessment-summary-policymakers-pdf

Jindal, R. (2012). Reducing poverty through carbon forestry? Impacts of the N'hambita community carbon project in Mozambique. *World Development*, *40*(10), 2123–2135.

Joshi, L. (2011). A community-based PES scheme for forest preservation and sediment control in Kulekhani, Nepal. In *FAO "Payments for ecosystem services and food security"* (pp. 198–203). FAO.

Khalumba, M., Wünscher, T., Wunder, S., Büdenbender, M., & Holm-Müller, K. (2014). Combining auctions and performance-based payments in a forest enrichment field trial in Western Kenya. *Conservation Biology*, *28*(3), 861–866.

Khatri, D. B. (2009). *Compromising the environment in Payments for Environmental Services? An institutional analysis of mechanisms for sharing hydroelectricity revenue in Kulekhani watershed, Nepal* [Master thesis, International Institute of Social Studies, The Hague, Netherlands, 51p].

Klimeka, S., Richter gen., Kemmermanna, A., Steinmann, H. H., Freese, J., & Isselstein, J. (2008). Rewarding farmers for delivering vascular plant diversity in managed grasslands: A transdisciplinary case-study approach. *Biological Conservation*, *141*, 2888–2897.

Kosoy, N., Martinez-Tuna, M., Muradian, R., Martinez-Alier, J. (2007). Payments for environmental services in watersheds: Insights from a comparative study of three cases in Central America. *Ecological Economics*, *61*, 446–455.

Krause, T., Collen, W., & Nicholas, K. A. (2013). Evaluating safeguards in a conservation incentive program: Participation, consent, and benefit sharing in indigenous communities of the Ecuadorian Amazon. *Ecology and Society*, *18*(4), 1.

LaRocco, G. L., & Deal, R. L. (2011). *Giving credit where credit Is due: Increasing landowner compensation for ecosystem services* (General Technical Report

PNW-GTR-842 April 2011). Pacific Northwest Research Station, Forest Service, United States Department of Agriculture.

Leathers, N., Harrington, L. (2000). Effectiveness of conservation reserve programs and land 'slippage' in southwestern Kansas. *Professional Geographer*, *52*, 83–93.

Leimona, B., Kelsey-Jack, B., Lusiana, B., & Pasha, R. (2009). *Designing a procurement auction for reducing sedimentation: A field experiment in Indonesia*. Economy and Environment Program for Southeast Asia research report no. 10. IDRC Regional Office for Southeast and East Asia, Singapore, SG.

Leimona, B., Pasha, R., & Rahadian, N. P. (2010). The livelihood impacts of incentive payments for watershed management in Cidanau watershed, West Java, Indonesia. In L. Tacconi, S. Mahanty, & H. Suich (Eds.), *Payments for environmental services, forest conservation and climate change livelihoods in the REDD?*. Edward Elgar Publishing.

Lewison, R. L., An, L., & Chen, X. (2017). Reframing the payments for ecosystem services framework in a coupled human and natural systems context: Strengthening the integration between ecological and human dimensions. *Ecosystem Health and Sustainability*, *3*(5). https://www.tandfonline.com/doi/full/10.1080/20964129.2017.1335931

Liu, J., Li, S., Ouyang, Z., Tam, C., & Chen, X. (2008). Ecological and socioeconomic effects of China's policies for ecosystem services. *Proceedings of the National Academy of Sciences*, *105*(28), 9477–9482. https://doi.org/10.1073/pnas.0706436105

Liu, J., Linderman, M., Ouyang, Z., An, L., Yang, J., & Zhang, H. (2001). Ecological degradation in protected areas: The case of Wolong nature reserve for giant pandas. *Science*, *292*(5514), 98–101.

Lobley, M., & Potter, C. (1998). Environmental Stewardship in UK agriculture: A comparison of the environmentally sensitive area programme and the Countryside Stewardship scheme in South East England. *Geoforum*, *29*(4), 413–432. https://doi.org/10.1016/S0016-7185(98)00019-0

Lopa, D., Mwanyoka, I., Jambiya, G., Massoud, T., Harrison, P., Ellis-Jones, M., Blomley, T., Leimona, B., van-Noordwijk, M., Burgess, N. D. (2012). Towards operational payments for water ecosystem services in Tanzania: A case study from the Uluguru Mountains. *Oryx*, *46*(1), 34–44.

Mayrand, K., & Paquin, M. (2004). *Payments for environmental services: A survey and assessment of current schemes*. Commission for Environmental Cooperation of North America.

McAfee, K., & Shapiro, E. N. (2010). Payments for ecosystem services in Mexico: Nature, neoliberalism, social movements, and the state. *Annals of the Association of American Geographers*, *100*(3), 579–599. https://doi.org/10.1080/00045601003794833

Millennium Ecosystem Assessment. (2003). *Ecosystems and human well-being: A framework for assessment*. Island Press.

Moreno-Sanchez, R., Maldonado, J. H., Wunder, S., & Borda-Almanza, C. (2012). Heterogeneous users and willingness to pay in an ongoing payment for watershed protection initiative in the Colombian Andes. *Ecological Economics*, *75*, 126–134.

Motallebi, M., Hoag, D. L., Tasdighi, A., Arabi, M., Osmond, D. L., & Boone, R. B. (2018). The impact of relative individual ecosystem demand on stacking ecosystem credit markets. *Ecosystem Services*, *29*, 137–144.

Muñoz-Piña, C., Guevara, A., Torres, J. M., & Braña, J. (2008). Paying for the hydrological services of Mexico's forests: Analysis, negotiations and results. *Ecological Economics*, *65*, 725–736.

Muradian, R., Corbera, E., Pascual, U., Kosoy, N., & May, P. H. (2010). Reconciling theory and practice: An alternative conceptual framework for understanding payments for environmental services. *Special Section - Payments for Environmental Services: Reconciling Theory and Practice*, *69*(6), 1202–1208. https://doi.org/10.1016/j.ecolecon .2009.11.006

Naeem, S., Ingram, J. C., Varga, A., Agardy, T., Barten, P., Bennett, G., Bloomgarden, E., Bremer, L. L., Burkill, P., Cattau, M., Ching, C., Colby, M., Cook, D. C., Costanza, R., DeClerck, F., Freund, C., Gartner, T., Goldman-Benner, R., Gunderson, J., ... & Wunder, S. (2015). Get the science right when paying for nature's services. *Science*, *347*(6227), 1206. https://doi.org/10.1126/science.aaa1403

Naidoo, R., Weaver, L. C., Diggle, R. W., Matongo, G., Stuart-Hill, G., & Thouless, C. (2016). Complementary benefits of tourism and hunting to communal conservancies in Namibia. *Conservation Biology*, *30*(3), 628–638.

Narloch, U, Pascual, U, & Drucker, A. G. (2011). Cost-effectiveness targeting under multiple conservation goals and equity considerations in the Andes. *Environmental Conservation*, *38*(4), 417–425.

Nelson, F. (2008). *Developing alternative frameworks for community-based conservation: Piloting payments for environmental services (PES) in Tanzania's Simanjiro plains* (p. 38). USAID TransLinks.

Osano, P., de Leeuw, J., & Said, M. (2012). Wildlife PES schemes and pastoral livelihoods in arid & semi-arid lands (ASALs) in Kenya. In Presentation at the Workshop Restoring Value to Grasslands, 7–10 May 2012, Brasilia, Brazil.

Pagiola, S. (2008). Payments for environmental services in Costa Rica. *Ecological Economics*, *65*(4), 712–724.

Pagiola, S. (2010). Desafíos y Oportunidades para el desarrollo de Pagos por Servicios Ambientales en el Sector Ganadero. In VI Agroforestry Congress on Cattle Sustainable Production, 28–30 September, 2010, Ciudad de Panama, Panamá.

Pagiola, S., Honey-Rosés, J., & Freire-González, J. (2017). *Assessing the permanence of land use change induced by payments for environmental services: Evidence from Nicaragua. Tropical Conservation Science*, *13*, 1940082920922676

Pagiola, S., Rios, A. R., & Arcenas, A. (2008). Can the poor participate in payments for environmental services? Lessons from the Silvopastoral Project in Nicaragua. *Environment and Development Economics*, *13*(3), 299–325.

Pagiola, S., Rios, A. R., & Arcenas, A. (2010). Poor household participation in payments for environmental services: Lessons from the Silvopastoral Project in Quindío, Colombia. *Environmental and Resource Economics*, *47*, 371–394.

Pereira, S. (2010). Payment for ecosystem services in the Amazon forest: How can conservation and development be reconciled? *The Journal of Environment & Development*, *19*(2), 171–190.

Perrot-Maître, D. (2005). Investing in protection of ecosystem services: A business opportunity for Vittel (Nestlé Waters), France. Paper presented at the ZEF–CIFOR Workshop on Payments for Environmental Services in Developed and Developing Countries, Titisee, Germany, 15–18 June 2005.

Perrot-Maître, D. (2006). *The Vittel payments for ecosystem services: A "perfect" PES case?* International Institute for Environment and Development.

Quintero, M., & Otero, W. (2006). *Mecanismo de financiación para promover Agricultura de Conservación con pequeños productores de la cuenca de la laguna de Fúquene. Su diseño, aplicación y beneficios*. Centro Internacional de la Papa.

Raes, L., Rengel, E., & Romero, J. (2012). *Inter-municipal cooperation in watershed conservation through the establishment of a regional water fund–FORAGUA–in Southern Ecuador.*

Robalino, J., Sandoval, C., Barton, D. N., Chacon, A., & Pfaff, A. (2015). Evaluating interactions of forest conservation policies on avoided deforestation. *PLOS ONE, 0124910,* 1–16.

Robertson, M., BenDor, T. K., Lave, R., Riggsbee, A., Ruhl, J., & Doyle, M. (2014). Stacking ecosystem services. *Frontiers in Ecology and the Environment, 12*(3), 186–193. https://doi.org/10.1890/110292

Robertson, N., & Wunder, S. (2005). *Fresh tracks in the forest: Assessing incipient payments for environmental services initiatives in Bolivia.* CIFOR.

Robinson, R. (1998). Community partnership in the Richtersveld National Park, South Africa. Paper presented in the Scandinavian Seminar College Workshop in Abidjan, Ivory Coast, November 9–11, 1998.

Schomers, S., & Matzdorf, B. (2013). Payments for ecosystem services: A review and comparison of developing and industrialized countries. *Ecosystem Services, 6,* 16–30.

Smith, S., Rowcroft, P., Everard, M., Couldrick, L., Reed, M., Rogers, H., Quick, T., Eves, C., & White, C. (2015). *Payments for ecosystem services: A best practice guide.* Defra. https://ecosystemsknowledge.net/sites/default/files/wp-content/uploads/PES%20Best%20Practice%20Guide%20-%202015%20edition.pdf

Sommerville, M., Jones, J. G. P., Rahajaharison, M., & Milner-Gulland, E. J. (2010). The role of fairness and benefit distribution in community-based payment for environmental services interventions: A case study from Menabe, Madagascar. *Ecological Economics, 69,* 1262–1271.

Song, C., Bilsborrow, R., Jagger, P., Zhang, Q., Chen, X., & Huang, Q. (2018). Rural household energy use and its determinants in China: How important are influences of payment for ecosystem services vs. Other factors? *Ecological Economics, 145,* 148–159. https://doi.org/10.1016/j.ecolecon.2017.08.028

Soriaga, R., & Annawi, D. (2010). The 'no-fire bonus' scheme in mountain province, cordillera administrative region, Philippines. In L. Tacconi, S. Mahanty, & H. Suich (Eds.), *Payments for environmental services, forest conservation and climate change livelihoods in the REDD?.* Edward Elgar Publishing.

Speiser, S., Bauer, K., & Villacreset, D. (2009). *Buenas Prácticas Conservación y Desarrollo: una experiencia de los Chachi en el Noroccidente Ecuatoriano.* GTZ. Available at http://www.unl.edu.ec/agropecuaria/wp-content/uploads/2012/03/sp-buenas_practicas-areas-protegidas-Chachi-ecuador1.pdf

Suyanto, S. (2007). *Lessons on the conditional tenure and RiverCare schemes in Sumberjaya, Indonesia: Conditionality in payment for environmental services. Insight: Notes from the field* (pp. 29–35). RECOFTC, Centro Mundial de Agrosilvicultura (ICRAF) y Winrock International India (WII).

Suyanto, S., Leimona, B., Permana, R. P., & Chandler, F. J. C. (2005). *Review of the development environmental services market in Indonesia.* World Agroforestry Center. Available at http://www.worldagroforestrycentre.org/sea.

The Nature Conservancy. (2022). Payment for ecosystem services. *Conservation Gateway.* https://www.conservationgateway.org/ConservationPractices/EcosystemServices/ValuationandPayments/PaymentforEcosystemServices/Pages/payment-ecosystem-service.aspx

Tipper, R. (2002). Helping indigenous farmers to participate in the international market for carbon services: The case of Scolel Té. In J. Bishop, S. Pagiola. (Eds.) *Selling forest environmental services: Market based mechanisms for conservation and development,*(pp. 223–234). Taylor & Francis.

To, P. X., Dressler, W. H., Mahanty, S., Pham, T. T., & Zingerli, C. (2012). The prospects for payment for ecosystem services (PES) in Vietnam: a look at three payment schemes. *Human Ecology*, *40*(2), 237–249.

Tuanmu, M.-N., Viña, A., Yang, W., Chen, X., Shortridge, A. M., & Liu, J. (Jack). (2016). Effects of payments for ecosystem services on wildlife habitat recovery. *Conservation Biology*, *30*, 827–835.

Uchida, E., Xu, J., & Rozelle, S. (2005). Grain for green: Cost-effectiveness and sustainability of China's conservation set-aside program. *Land Economics*, *81*(2), 247–264.

UNFCCC. (2016). *The REDD desk*. https://theredddesk.org/what-redd

United Nations. (2016). *The 2030 agenda for sustainable development (17 sustainable development goals or SDGs)*. https://sustainabledevelopment.un.org/sdgs

United States Department of Agriculture, & Farm Service Agency. (2019). *Conservation reserve program statistics*. https://www.fsa.usda.gov/programs-and-services/ conservation-programs/reports-and-statistics/conservation-reserve-program-statistics/ index

Weaver, C., & Petersen, T. (2008). Namibia communal area conservancies. *Best Practices in Sustainable Hunting*, 48–52.

Wendland, K., & Suárez, L. (2009). *Incentivos a cambio de servicios ambientales y colectividad de la tenencia de la tierra: Lecciones de Ecuador e Indonesia*. USAID Land Tenure Center, Tenure Brief no. 9.

World Resources Institute. (2009). *Stacking payments for ecosystem services* (WRI FACT SHEET). World Resources Institute. https://www.wri.org/. https://pdf.wri .org/factsheets/factsheet_stacking_payments_for_ecosystem_services.pdf?_ga=2 .255694337.867138834.1542689267-609383282.1542689267

Wunder, S. (2005). *Payments for environmental services: Some nuts and bolts*. CIFOR. https://books.google.com/books?id=L6YsAQAAMAAJ

Wunder, S. (2008). Payments for environmental services and the poor: Concepts and preliminary evidence. *Environment and Development Economics*, *13*(03), 279–297.

Wunder, S. (2015). Revisiting the concept of payments for environmental services. *Ecological Economics*, *117*, 234–243. https://doi.org/10.1016/j.ecolecon.2014.08.016

Wunder, S, & Albán, M. (2008). Decentralized payments for environmental services: The cases of Pimampiro and PROFAFOR in Ecuador. *Ecological Economics*, *65*, 685–698.

Wunder, S., & Wertz-Kanounnikoff, S. (2009). Payments for ecosystem services: A new way of conserving biodiversity in forests. *Journal of Sustainable Forestry*, *28*(3–5), 576–596.

Wunder, S., Brouwer, R., Engel, S., Ezzine-de-Blas, D., Muradian, R., Pascual, U., & Pinto, R. (2018). From principles to practice in paying for nature's services. *Nature Sustainability*, *1*(3), 145–150. https://doi.org/10.1038/s41893-018-0036-x

Wunder, S., Engel, S., & Pagiola, S. (2008). Taking stock: A comparative analysis of payments for environmental services programs in developed and developing countries. *Ecological Economics*, *65*(4), 834–852. https://doi.org/10.1016/j.ecolecon.2008.03.010

Xu, J., Tao, R., Xu, Z., & Bennett, M. T. (2010). China's sloping land conversion program: Does expansion equal success? *Land Economics*, *86*(2), 219–244.

Yatich, T., Said, M., Swallow, B., & Sononka, J. (2008). Kitengela wildlife lease programme: Is it realistic, conditional, pro-poor and voluntary? In East and Southern Africa Katoomba Group Regional Workshop on Taking stock and charting a way forward: Payments for Ecosystem Services in Africa, 16–17 September 2008, Dar-es-salaam and Morogoro, Tanzania.

3 Concurrent green initiatives in the USA

With several essential issues in concurrent green initiatives—concurrent payments for environmental services in particular—identified and reviewed, this chapter turns to two specific concurrent green initiatives in the USA. On the one hand, we intend to show empirical evidence for spillover effects between the two initiatives. On the other hand, we provide some technical details (e.g., models, procedures) for how we arrive at the conclusion for the purpose of classroom teaching or education.

3.1 Two major green initiatives

The Conservation Reserve Program (CRP), authorized by the 1985 Farm Security Act and operated by the US Department of Agriculture (USDA hereafter), aims to retire environmentally sensitive land from agricultural production for 10–15 years (Riley, 2004). Such sensitive land is mainly located in highly erodible places. The central agri-environmental policy in the USA before 2002 used funds to pay retired farmers and low-income farmers (Claassen et al., 2008). The CRP has successfully reached its aims of preserving soil, water, and wildlife. For instance, the CRP has led to decreased cultivated acreage from 26% of the land area to 8% in Goodwin Creek, Mississippi. Perennial grasses established through the CRP have significantly improved infiltration and soil quality relative to conventional cropping systems at the Mark Twain Lake/Salt River Basin, Missouri. Substantial cropland area has been converted to grass or forest through the CRP at Yalobusha River and Topashaw Creek, Mississippi (Richardson et al., 2008). The CRP also benefits wildlife and fish (Gray & Teels, 2006).

The Environmental Quality Incentives Program (EQIP) was created in 1996 (via the 1996 Farm Bill) by consolidating several programs related to cropland and grazing land. Administered by Natural Resources Conservation Service (NRCS), the EQIP intends to pay agricultural producers to adopt environmentally friendly practices on their farmlands—i.e., lands that remain in production. So the EQIP is a working-land program and has cost-sharing for specific conservation practices (Ogg & Keith, 2002). Recent years, however, have witnessed increases in funding for working-land programs (e.g., EQIP) relative to land retirement programs (e.g., CRP). EQIP and CRP are concurrent payments for environmental

DOI: 10.4324/9781003290292-3

services (PES) programs according to the 2018 Farm Bill amendments, which explicitly "allow[s] land enrolled in CRP during the last year of the CRP contract to be enrolled in the Environmental Quality Incentives Program" (Federal register, 2019). In this situation, there exists a substantial potential for spillover effects between CRP and EQIP.

3.2 Potential spillover effects between CRP and EQIP

As CRP and EQIP are large US agri-environmental programs, we searched the Web of Science under the keyword in this format "(((Conservation Reserve Program or CRP) and (Environmental Quality Incentives Program or EQIP))" while selecting "All fields" (this is the most comprehensive choice compared with other alternatives such as "Title" and "Topic"). Then, we reviewed the abstracts and keywords of all the selected papers from this search: if both "Conservation Reserve Program" (or its acronym CRP) and "Environmental Quality Incentives Program" (or its acronym EQIP) occur in the abstract or keyword list, we consider it a potential paper addressing EQIP–CRP spillover effects. Otherwise, we skip it. Then, for all possible papers, we downloaded and read them in search of evidence of spillover effects.

We found 76 papers with publication dates ranging from 2000 to 2021 as of December 30, 2021. Of these 76 papers, 16 met a high standard for potentially addressing spillover effects. Out of the 16 articles, three suggest spillover effects—at least concurrency—between multiple green initiatives. First, Mishra and Khanal (2013) mention that a landowner can explicitly enroll in both programs, implying that EQIP and CRP meet our concurrent PES definition. Another paper indicates explicitly that in the Topashaw Canal watershed, USA, "interest in and sign-up for CRP began again in 1997 but dwindled to less than 2000 ha (4,942 ac) with payments of S20,000 per year once EQIP was initiated in 2002" (Wilson et al., 2008), indicating an offsetting spillover effect from EQIP to CRP. The third paper (Rossi et al., 2021) states that "Additional field experiments could reveal if these stated beliefs reflect the true motivations for farmers' enrollment in both programs," which implied that farmers have enrolled in both programs. We also found an implicit statement: "working-land and land retirement programs play complementary roles to reduce the environmental consequences of agricultural production" (Lambert et al., 2007), yet we found no discussion of their complementarity. However, no systematic work has been devoted to exploring such spillover effects.

Below are all 16 papers that potentially address EQIP–CRP spillover effects:

Claassen, R., Cattaneo, A., & Johansson, R. (2008). Cost-effective design of agri-environmental payment programs: U.S. experience in theory and practice. *Ecological Economics*, *65*(4), 737–752.

Rossi, G. D., Hecht, J. S., & Zia, A. (2021). A mixed-methods analysis for improving farmer participation in agri-environmental payments for ecosystem services in Vermont, USA. *Ecosystem Services*, *47*, 101223.

Frimpong, E. A., Lee, J. G., & Ross-Davis, A. L. (2007). Floodplain influence on the cost of riparian buffers and implications for conservation programs. *Journal of Soil and Water Conservation, 62*(1), 33–39.

Gray, R. L., & Teels, B. M. (2006). Wildlife and Fish Conservation Through the Farm Bill. *Wildlife Society Bulletin, 34*(4), 906–913.

Hess, G. R., Campbell, C. L., Fiscus, D. A., Hellkamp, A. S., McQuaid, B. F., Munster, M. J., Peck, S. L., & Shafer, S. R. (2000). A Conceptual Model and Indicators for Assessing the Ecological Condition of Agricultural Lands. *Journal of Environmental Quality, 29*(3), 728–737.

Lambert, D. M., Sullivan, P., Claassen, R., & Foreman, L. (2007). Profiles of US farm households adopting conservation-compatible practices. *Land Use Policy, 24*(1), 72–88.

Mishra, A. K., & Khanal, A. R. (2013). Is participation in agri-environmental programs affected by liquidity and solvency? *Land Use Policy, 35*, 163–170.

Medina, G., Isley, C., & Arbuckle, J. (2021). Promoting sustainable agriculture: Iowa stakeholders' perspectives on the US Farm Bill conservation programs. *Environment, Development and Sustainability, 23*(1), 173–194.

Mutandwa, E., Grala, R. K., Grado, S. C., & Munn, I. A. (2016). Family Forest Owners' Familiarity with Conservation Programs in Mississippi, USA. *Small-Scale Forestry, 15*(3), 303–319.

Ogg, C. W., & Keith, G. A. (2002). New Federal Support for Priority Watershed Management Needs. *JAWRA Journal of the American Water Resources Association, 38*(2), 577–586.

Reimer, A. P., & Prokopy, L. S. (2014). Farmer Participation in U.S. Farm Bill Conservation Programs. *Environmental Management, 53*(2), 318–332.

Richardson, C. W., Bucks, D. A., & Sadler, E. J. (2008). The Conservation Effects Assessment Project benchmark watersheds: Synthesis of preliminary findings. *Journal of Soil and Water Conservation, 63*(6), 590–604.

Riley, T. Z. (2004). Private-land habitat opportunities for prairie grouse through federal conservation programs. *Wildlife Society Bulletin, 32*(1), 83–91.

Tumeo, M. A., Mauriello, D. A., Sadeghi, A. M., & Meekhof, R. (2000). Case Studies on the Application of Adaptive Risk Analysis to USDA's Resource Conservation Programs. In Y. Y. Haimes & R. E. Steuer (Eds.), *Research and practice in multiple criteria decision making* (pp. 492–509). Springer.

Tyndall, J. (2021). Prairie and tree planting tool—PT2 (1.0): A conservation decision support tool for Iowa, USA. Agroforestry Systems, 1–16.

Wilson, G. V., Shields, F. D., Bingner, R. L., Reid-Rhoades, P., DiCarlo, D. A., & Dabney, S. M. (2008). Conservation practices and gully erosion contributions in the Topashaw Canal watershed. *Journal of Soil and Water Conservation, 63*(6), 420–429.

3.3 Empirical data collection and analysis

We obtained EQIP data in 2018 on a county basis from the USDA (USDA Farm Production and Conservation Business Center, 2020). We downloaded county-level CRP data in 2018 from the USDA Farm Service CRP program and statistics reporting portal (https://www.fsa.usda.gov/programs-and-services/conservation-programs/reports-and-statistics/conservation-reserve-program-statistics/index) on April 6, 2020. The income data were downloaded from the US

Census—SAIPE (Small Area Income and Poverty Estimates) and the related links (https://www.census.gov/data/datasets/2018/demo/saipe/2018-state-and-county .html and https://www.census.gov/programs-surveys/saipe.html). The farmland data were downloaded from the USDA—Farm Service Agency (https://www.fsa .usda.gov/news-room/efoia/electronic-reading-room/frequently-requested-infor-mation/crop-acreage-data/index). The population data were downloaded from the US Census Bureau (https://www.census.gov/data/tables/time-series/demo/popest /2010s-counties-total.html).

After merging the datasets by county, we generated a dataset that contains the following variables: CRP2018 (y for area enrolled in CRP; acres), EQIP_ Area (x_1 for contracted land in EQIP; acres), Farm_Area (x_2 for total county farmland; acres), M_HH_inc (x_3 for county median household income in 2018; $), and Pop2018 ($x_4$ for county population in 2018). As a preliminary initiative to handle spatial autocorrelation in the dataset, we first randomly selected 25% of the data out of 3,108 records, resulting in a dataset of 730 records for data analysis. According to the United Nations' Sustainable Livelihoods Framework, human, social, natural, physical, and financial capitals possessed by an entity (e.g., farm, household, community) play a crucially important role in relevant livelihood decisions. Using the acres of EQIP enrollment as dependent variable (y), we explain its variability using a set of variables that represent such capitals: the acres of CRP enrollment (X_1), total farmland (X_2, acres), median household income (X_3), and population size (X_4). The multivariate linear regression takes the following form (Equation 3.1):

$$y = b_0 + b_1 X_1 + \sum_{i=2}^{4} b_i X_i + e \qquad (3.1)$$

where b_0 is the intercept, b_1 is the coefficient of X_1 (contracted land in EQIP; acres), and b_i is the coefficient of the three control variables X_i (i=2, 3, and 4) that contribute to explaining the variability in the dependent variable (land enrolled in CRP; acres). As shown later, the results indicate that under the control of county-level farmland area, income, and population size, EQIP land had a negative impact on CRP enrollment—each acre of EQIP land caused a loss of 0.28 ($p < 0.0001$) acre in CRP enrollment.

We further employed the eigenvector spatial filtering (ESF) method to handle potential biases in parameter estimates due to spatial autocorrelation (Chun, 2008; Griffith, 2000). Spatial autocorrelation refers to a situation where units that are geographically close to one another may have more similar values than those that are far apart, which is also known as Tobler's first law of geography (Tobler, 1970). If this type of autocorrelation is present in a regression model (e.g., in its residuals), then it violates a fundamental assumption in standard statistical analysis: regression residuals should be independent and identically distributed (i.i.d.). The violation may give rise to biased parameter estimates, e.g., an increase in type I error and falsely rejecting the null hypothesis of no effect.

Employing the ESF method, we tested various neighborhood sizes from the 1st- to the 20th-order Queen's neighborhood as we do not know precisely at what spatial scale(s) the residuals are spatially autocorrelated. At each neighborhood size, we generated the corresponding spatial weights matrix.

Following the relevant literature (An et al., 2016; Chun et al., 2016), we calculated the eigenvalues (ranked in descending order) and the corresponding eigenvectors under each neighborhood size (i.e., 1, 2 ... up to 20). According to eigenvector selection literature (Hughes & Haran, 2013; Pace et al., 2013), a relatively small subset of top eigenvectors (e.g., top 50–100 for a dataset with 2,500 records; Hughes & Haran, 2013) should suffice as regressors for filtering out spatial autocorrelation. We name this procedure the "top k method" for illustration purposes, where it is essential to determine the value of k. One way to choose k is to select the top k eigenvectors corresponding to standardized eigenvalues greater than 0.7 (we name it the 0.7 rule; Hughes & Haran, 2013).

Alternatively, the top k eigenvectors can be determined by the "0.25 rule" (as described for illustration convenience; Chun et al., 2016), and the number thus chosen should be more than the subset defined by the 0.7 rule. Note that the 0.25 rule states that k can be determined if $EV_k/EV_{max} \geq 0.25$ for positive spatial autocorrelation, where EV_{max} is the largest eigenvalue among all n eigenvalues (Chun et al., 2016). We show at each neighborhood size, the maximum eigenvalue, a quarter of the maximum eigenvalue (i.e., 0.25× maximum eigenvalue), and the number of eigenvectors with their eigenvalue greater than 0.25× maximum eigenvalue (Table 3.1). For instance, at neighborhood=2 (the second-order neighborhood is chosen for eigenvector calculation), there are 290 eigenvectors with eigenvalues greater than 5.23 (here 5.23=0.25 × 20.91, where 20.91 is the maximum eigenvalue).

Therefore, we picked up the top k eigenvectors for the regression based on the 0.25 rule. The regression model is shown in Equation 3.2:

$$y = b_0 + b_1 x_1 + b_2 x_2 + b_3 x_3 + b_4 x_4 + \sum_{g=1}^{k} c_g CEV_g + e \qquad (3.2)$$

Based on Equation 3.2, we calculated regression residuals, Moran's I value, and the associated Z score at each of the 20 neighborhood choices. Following Chun et al. (2016), we chose the appropriate model (corresponding to a specific neighborhood size) that (1) reduces spatial autocorrelation to an acceptable level (e.g., |z| is less than 1.64 for alpha = 0.10) and (2) has the best (or close to the best) model fit in terms of, e.g., minimized AIC or maximized adjusted R^2. The first rule prevails if these two rules cannot be satisfied simultaneously. When multiple models (each for a unique neighborhood size) satisfy these two rules, we choose the one with fewer eigenvectors for higher degrees of freedom.

The results based on Equation 3.2 indicate that at the tenth order, the spatial autocorrelation of residuals was nearly removed with $|z| = 0.54$ (Table 3.2). At this neighborhood size (i.e., the tenth order) with the least $|z|$ score, k was determined

Table 3.1 The number of eigenvectors selected at each neighborhood size

Order	1	2	3	4	5	6	7	8	9	10
Max value	6.71	20.91	43.49	74.14	111.73	155.25	204.08	257.76	315.07	375.29
0.25 of max	1.68	5.23	10.87	18.53	27.93	38.81	51.02	64.44	78.77	93.82
Top # selected	>500	290	145	87	57	40	30	23	19	15

Order	11	12	13	14	15	16	17	18	19	20
Max value	437.38	500.07	562.59	624.13	683.70	740.56	793.92	843.20	887.87	927.47
0.25 of max	109.35	125.02	140.65	156.03	170.93	185.14	198.48	210.80	221.97	231.87
The last value	13	10	9	8	7	7	6	5	4	4

Table 3.2 Spatial autocorrelation of regression residuals for CRP and EQIP, USA

Neighborhood order	Under normality assumption			Under randomization assumption		
	Moran's I[a]	p-value	Z score	Moran's I	p-value	Z score
1	−0.0946	1.0000	−8.9821	−0.0946	1.0000	−9.0694
2	−0.0515	1.0000	−8.6582	−0.0515	1.0000	−8.7351
3	−0.0137	0.9994	−3.2550	−0.0137	0.9995	−3.2843
4	−0.0270	1.0000	−8.5057	−0.0270	1.0000	−8.5746
5	−0.0253	1.0000	−9.8717	−0.0253	1.0000	−9.9512
6	−0.0130	1.0000	−5.9994	−0.0130	1.0000	−6.0484
7	−0.0044	0.9877	−2.2476	−0.0044	0.9883	−2.2657
8	0.0070	0.0000	4.6093	0.0070	0.0000	4.6472
9	0.0066	0.0000	4.9065	0.0066	0.0000	4.9455
10	−0.0010	0.7033	−0.5339	−0.0010	0.7047	−0.5380
11	−0.0086	1.0000	−7.2702	−0.0086	1.0000	−7.3251
12	−0.0113	1.0000	−10.5581	−0.0113	1.0000	−10.6381
13	−0.0166	1.0000	−17.0326	−0.0166	1.0000	−17.1610
14	−0.0185	1.0000	−20.6911	−0.0185	1.0000	−20.8460
15	−0.0190	1.0000	−22.9488	−0.0190	1.0000	−23.1202
16	−0.0174	1.0000	−22.5684	−0.0174	1.0000	−22.7360
17	−0.0137	1.0000	−18.9725	−0.0137	1.0000	−19.1124
18	−0.0088	1.0000	−12.9496	−0.0088	1.0000	−13.0441
19	−0.0046	1.0000	−6.9333	−0.0046	1.0000	−6.9840
20	−0.0015	0.9829	−2.1183	−0.0015	0.9836	−2.1336

Notes: [a] When calculating the spatial weights matrix, the few records (counties) in California were dropped as most counties in California did not have both CRP and EQIP implemented simultaneously, leaving few scattered counties in our dataset. Also in order to calculate Moran's I, counties without a residual were assigned the average of residuals of all residuals.

to be 15 based on the above 0.25 rule. Using these top 15 eigenvectors as spatial filters, the area of EQIP land had a negative coefficient of −0.2242 ($p < 0.0001$) (Table 3.3, the second model).

To examine whether the model based on a subset of 730 records (Table 3.3) can reduce the spatial autocorrelation to an acceptable level, we also calculated the Moran's I value of this model at four neighborhood levels, i.e., the 5th, 10th, 15th, and 20th. It turns out the residuals were still quite spatially autocorrected except at the 15th level ($z = 1.5672$; Table 3.4). This suggests that the subsampling method may not effectively reduce spatial autocorrelation.

We used the stepwise selection method to choose spatial filters (Chun et al., 2016; Chun & Griffith, 2011; Griffith, 2000) to verify the above results. Under this method, we used stepwise regression to select a subset of s significant eigenvectors (at alpha = 0.10) out of the top k eigenvectors. In our regression model (Equation 3.3), the top k eigenvectors—candidate spatial filters that were chosen based on the 0.25 rule—entered the stepwise procedure (note that X_1 through X_4 were forced to be included). These s eigenvectors were then used as spatial filters in the regression model that corresponds to a specific neighborhood size:

Table 3.3 Regression results for the CRP and EQIP, USA

Variable	Model with n = 730 (669 used)				Model with ESF (tenth-order neighborhood)[a]			
	Coefficient	*t-score*	*p-value*	*Variance inflation*	*Coefficient*	*t-score*	*p-value*	*Variance inflation*
Intercept	3713.7310	1.64	0.1010	0	8017.8357	5.18	<0.0001	0
EQIP_Area	−0.2823	−5.39	<0.0001	1.4413	−0.2242	−7.72	<0.0001	1.3648
Farm_Area	0.0215	12.14	<0.0001	1.4628	0.0174	13.67	<0.0001	2.0796
M_HH_Inc	0.0061	0.14	0.8903	1.2393	−0.0563	−1.95	0.0509	1.4232
Pop2018	−0.0108	−2.75	0.0061	1.2589	−0.0033	−2.11	0.0352	1.1620
V1	N/A	N/A	N/A	N/A	210474	9.26	<0.0001	1.7010
V2	N/A	N/A	N/A	N/A	1087.7090	0.06	0.9523	1.0793
...								
V14	N/A	N/A	N/A	N/A	−49285	−2.82	0.0048	1.0208
V15	N/A	N/A	N/A	N/A	9397.7823	0.52	0.6039	1.0417
Model fit	$R^2=0.2064$, Adjusted $R^2 = 0.2016$				$R^2=0.2253$, Adjusted $R^2=0.2201$			

Note: [a] For eigenvectors, we only show the first two and last two for brevity. We chose to show results for regression with a subset of data ($n=730$) and with eigenvector spatial filtering (ESF; at the tenth-order neighborhood) for the case of CRP and EQIP, USA.

Table 3.4 Spatial autocorrelation of regression residuals in the baseline model

Neighborhood order	Under normality assumption			Under randomization assumption		
	Moran's I	p-value	Z score	Moran's I	p-value	Z score
5	0.0450	0.0000	17.9028	0.0450	0.0000	18.2122
10	0.0140	0.0000	11.3534	0.0140	0.0000	11.5468
15	0.0009	0.0616	1.5417	0.0009	0.0585	1.5672
20	0.0034	0.0000	6.5202	0.0034	0.0000	6.6235

Notes: The results are based on the baseline model with a subset of $n=730$ records but no eigenvectors, i.e., the first model in Table 3.2.

Table 3.5 Regression results with the ESFs selected by stepwise regression

Variable	Coefficient	t-score	p-value	Variance inflation
Intercept	7,447.4780	5.23	<0.0001	0
EQIP_Area	−0.2240	−7.90	<0.0001	1.3000
Farm_Area	0.0172	14.02	<0.0001	1.9338
M_HH_Inc	−0.0440	−1.65	0.0981	1.2138
Pop2018	−0.0035	−2.26	0.0238	1.1413
V1	212,815	9.59	<0.0001	1.6214
V6	−75,696	−4.27	<0.0001	1.0488
V9	−66,165	−3.71	0.0002	1.0308
V10	62,279	3.60	0.0003	1.0308
V11	−96,370	−5.43	<0.0001	1.0373
V12	45,740	2.58	0.0099	1.0247
V14	−49,891	−2.86	0.0042	1.0182

$$y = b_0 + b_1x_1 + b_2x_2 + b_3x_3 + b_4x_4 + \sum_{g=1}^{s} c_g CEV_g + e \tag{3.3}$$

where x_i ($i=1, 2, 3,$ and 4) are the four predictor variables defined in Equation 3.1, and CEV_g and c_g ($g=1, 2, 3, \dots s$; $s \leq k$) are the eigenvectors that are chosen as spatial filters and the associated coefficients, respectively. Note that the s chosen eigenvectors are not necessarily the top s eigenvectors; therefore, CEV_g in Equation 3.3 could differ from CEV_g in Equation 3.2.

The regression results from Equation 3.3 also indicate that the tenth-order neighborhood is also acceptable with $|z| = 1.4777$ ($p=0.0697$, not shown in a table), confirming the outcome regarding the tenth-order neighborhood by Equation 3.2. For the brevity purpose, we skip the Moran's I and other statistics for this method as we did for the top k method (Table 3.1).

The coefficient for the area of EQIP land (EQIP_Area) is −0.2240 ($p < 0.0001$; Table 3.5), slightly different from that from Equation 3.2 (−0.2242; $p < 0.0001$; Table 3.3). Later we adopted the average of the two coefficients when calculating the impacts of EQIP on CRP: coefficient $= [(−0.2242)+(−0.2240)]/2 = −0.2241,$

rounded to –0.22. When calculating the average coefficient, we did not include the coefficient from Equation 1, i.e., –0.2803 (i.e., the one from the $n = 730$ sample; Table 3.3), simply because we preferred a conservative estimate.

3.4 Potential reasons for the negative spillover effects

There is an offsetting spillover effect from EQIP to CRP, which can be explained as follows. First, we regard land scarcity as a top influential variable. The positive coefficient of total farmland (0.0172 with $p < 0.0001$; Table 3.5) indicates more enrollment in CRP in counties with more farmland. Second, land-use competition may also—at least partially—account for this offsetting impact. Both CRP and EQIP target farmland, sharing goals to preserve water, soil, and wildlife habitat—so increases in enrollment of one program may lead to decreases in that of the other.

Last but not least, facing two choices of CRP and EQIP that are competitive in many instances, landowners choose the more profitable one. As CRP participants must retire the enrolled land, the land then has no (or very little) agricultural income. The EQIP, instead of retiring the land, provides money to landowners for whatever environmentally beneficial practices they adopt. This operation implies that landowners still receive agricultural income. Furthermore, the EQIP pay rate was higher than that of the CRP, offering an additional incentive for landowners to participate in the EQIP rather than the CRP. All these factors may contribute to the declining CRP enrollment trend since 2007 (Figure 3.1) and explain why the 2018 CRP enrollment was far below the cap designated by the 2018 Farm Bill.

In 2018, 22 million acres of land were enrolled in CRP, far less than the cap of 27 million acres established by the 2018 Farm Bill (USDA Farm Service Agency,

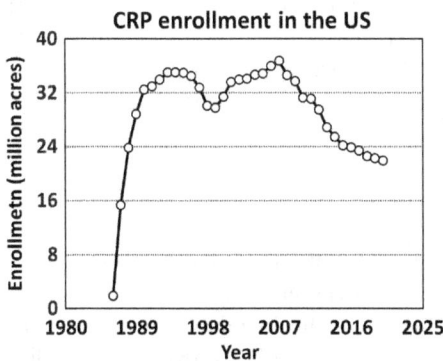

Figure 3.1 Dynamics of CRP-enrolled acres in the USA between 1986 and 2018 (Data source is USDA Farm Production and Conservation Business Center, Economics and Policy Analysis Division, Data Services Branch, County-level CRP, and EQIP dataset in the USA, 2020).

Table 3.6 Potential loss of CRP land area due to EQIP enrollment over time

Year	EQIP (million acres)[a]	Lost CRP (million acres)[b]	Soil erosion loss (billion)[c]	Carbon loss (million)	Cars back (million)
2009	23.1752	5.0985	2.0858	11.3558	2.0858
2010	24.1148	5.3053	2.1703	11.8163	2.1703
2011	22.4588	4.9409	2.0213	11.0048	2.0213
2012	24.3016	5.3464	2.1871	11.9078	2.1871
2013	24.4184	5.3721	2.1977	11.9650	2.1977
2014	19.4651	4.2823	1.7519	9.5379	1.7519
2015	18.6048	4.0930	1.6744	9.1163	1.6744
2016	15.7342	3.4615	1.4161	7.7098	1.4161
2017	17.0746	3.7564	1.5367	8.3665	1.5367
2018	17.7341	3.9015	1.5961	8.6897	1.5961
2019	18.0222	3.9649	1.6220	8.8309	1.6220
2020	17.6730	3.8881	1.5906	8.6598	1.5906
Average	20.2314	4.4509	1.8208	9.9134	1.8208

Notes:
[a] Data source: https://www.nrcs.usda.gov/Internet/NRCS_RCA/reports/fb08_cp_eqip.html.
[b] The lost CRP areas are estimated based on our modeled correlation, which is 22% of the total EQIP area. The data come from the dataset from the USDA (USDA Farm Production and Conservation Business Center, 2020).
[c] The news release from the USDA (USDA Farm Service Agency, 2019).

2019, p. 9). It is reported that all CRP land can generate a large amount of eco-logical benefits, including reduction of soil erosion (accumulative number) at the magnitude of 9 billion tons and sequestration of 49 million tons of carbon dioxide (equal to taking 9 million cars off the roads) (USDA Farm Service Agency, 2019). Our data show that between 2009 and 2020, the EQIP land was 20.2314 million acres (8.1874 million hectares) on average, which may have reduced CRP land by 4.4509 million acres (1.8012 million hectares) or 20.23% (100% × 4.4509 million/22 million) total CRP land. Given such data, the loss of CRP land due to its competitor EQIP (4.4509 million acres) is equivalent to increasing soil erosion by 1.8208 billion tons and carbon dioxide release by 9.9134 million tons. The increased release of greenhouse gas alone would be equivalent to putting 1.8208 million cars back on roads (Table 3.6).

3.5 Area-based conservation experiment

As discussed in Section 1.4, the conservation community has long established and maintained protected areas. Protected areas are considered an essential type of green initiative, which is credited to be the foundation of biodiversity conservation. The conservation community has recently started recommending area-based conservation measures for conservation purposes (Jonas et al., 2014; Maxwell et al., 2020).

We experimented to explore whether and how we can leverage the spillover effects to save costs while maintaining the total acres enrolled in CRP and EQIP.

We first convert various proportions of EQIP land located in areas eligible for both programs back to the CRP. The rationale is that some landowners may quit their land from the CRP—though more ecologically appropriate under this program—but enroll such land in the EQIP for its higher pay rate. The experiment begins with 22.0 and 18.02 million acres of CRP and EQIP enrollment in 2019, respectively, which stands as the baseline. Based on our finding in Section 3.3, each acre of EQIP land may lead to a reduction of 0.22 acre in CRP land, which suggests that 20.23 million EQIP lands (the average from 2009 to 2020) should have reduced CRP enrollment at the magnitude of 20.23 × 0.22 = 4.5 million acres.

Next, we consider five scenarios: zero (pre-pandemic, baseline), 25% (4.5 × 25% = 1.125 million acres reallocated from EQIP to CRP), 50%, 75%, and up to 100% restoration (all 4.5 million acres reallocated from EQIP to CRP; Table 3.7). As we move the same amount of EQIP acres to the CRP, the total acres in both programs remain the same, but the amount of total payment declines simply because the pay rate of CRP ($76.36/acre) is lower in comparison to that of EQIP ($137.98/acre). The results show that 2–7% of the total expenses would have been saved while still meeting the goal of constant acreage of both EQIP and CRP (Table 3.7).

If there are no spillover effects, the green efforts are invested in Figure 3.2A. However, our data analysis found that in areas with both CRP and EQIP eligible, a certain amount of land (22%, the dotted oval in Figure 3.2B), which is best for the CRP, has switched to the EQIP because of its higher pay rates and other benefits (Figure 3.2B). As a result, the total amount of land for CRP and EQIP remains unchanged, but the total ecological benefits will likely decrease, and the total payments for both programs will increase.

What is the application of this finding to landscape design and engineering? As we did in the China case (Section 6.6), we suggest changing the enrollment rules to some degree, such that some EQIP efforts in the middle section—the dotted circle—can be reallocated to EQIP-only areas (Figure 3.2C). As a result, the total area of both CRP and EQIP enrollment remains unchanged, but we get the effort (in the middle area) reallocated to the best ecological benefits. The other benefit is that through such a reshuffle of green efforts, we can allow a 2–7% budget cut for the whole USA but still keep the total area of both CRP and EQIP unchanged. Budget readjustments are particularly important at times of crisis, such as COVID-19.

Appendix: Moran's I calculation

We present a detailed description of how to calculate the Moran's I value at the various neighborhood definitions that are shown in Table 3.2. For data and related code, visit our website http://www.complexties.org/book/green_initiative, and go to this subfolder EQIP-CRP-data/Moran-I/. Readers without interest in such detail scan skip this section.

Table 3.7 Payments saved due to policy redesign between CRP and EQIP

| | Pre-pandemic | | During or post-pandemic | | | | | | | |
| | Baseline[a] | | 25% restoration[b] (1.125 m. acre) | | 50% restoration (2.25 m. acre) | | 75% restoration (3.375 m. acre) | | 100% restoration (4.5 m. acre) | |
	Area	Pay	Area	Pay	Area	Pay	Area	Pay	Area	Pay
CRP	22.00	1,680	23.125	1,766	24.25	1,852	25.375	1,938	26.5	2,024
EQIP[c]	18.02	2,486	16.895	2,331	15.77	2,176	14.645	2,021	13.52	1,865
Total	40.02	4,166	40.02	4,097	40.02	4,028	40.02	3,958	40.02	3,889
Change	N/A	N/A	0	−69	0	−139	0	−208		−277
Change %	N/A	N/A	0	−2%	0	−3%	0	−5%		−7%

Notes:
[a] Based on 2019 data, we consider both area (million acres) and payment (million dollars).
[b] The base of restoration is 4.5 million acres (total lost CRP land due to EQIP).
[c] EQIP data from https://www.nrcs.usda.gov/Internet/NRCS_RCA/reports/fb08_cp_eqip.html.

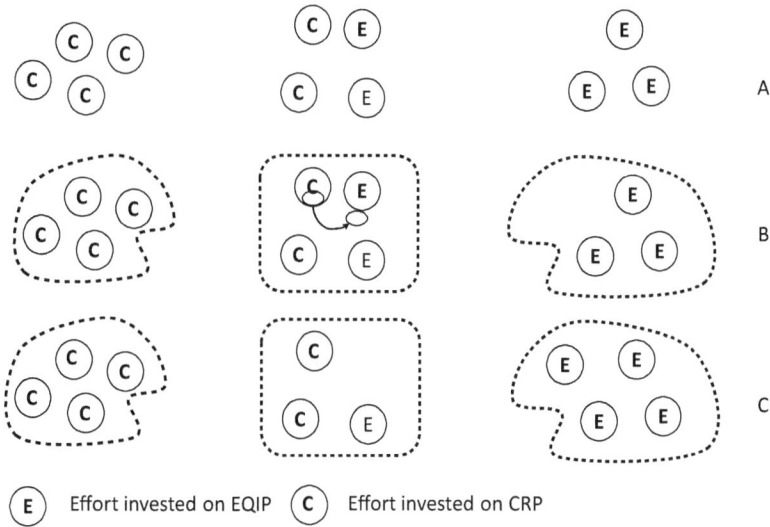

Figure 3.2 Initiative efforts invested in EQIP and CRP if (A) there are no spillover effects, (B) there is a spillover effect, and (C) the influenced EQIP effort is relocated to EQIP-only areas. The left, middle, and right dotted areas represent CRP eligible only areas, CRP and EQIP simultaneously eligible areas, and EQIP eligible only areas.

Step 1: Data preparation

1.1 Create folders named **NonSpatial, Results, and Shapefiles**; move **Out _no_EV.csv** into the **Nonspatial** folder.

1.2 Import **out_data_1.csv** into ArcMap, join with county layer by FIPS_co. Extract the matching records as a new shapefile named **county_outdata .shp** under the Shapefiles folder.

1.3 Fill in missing residuals in each table from average of all residuals (Step 1 in MoranI.R)

Step 2: Calculate the neighborhood for each order specified in the table names (Step 2 in MoranI.R)

Step 3: Calculate Moran's I.

   We calculate Moran's I values for regression residuals and related statistics (Step 3 in MoranI.R); two tables are found in the Results folder (**MoranI _nor.csv** for Moran's I under normality and **MoranI_ran.csv** for Moran's I under randomization).

Step 4: Prepare a table for non-spatial model (Step 4 in MoranI.R)

   4.1: merge the non-spatial table with the full county layer (3,106 counties)

   4.2: fill in missing residuals in each table from an average of all residuals

Step 5: Calculate Moran's I for regression residuals and related statistics (Step 5 in MoranI.R); two tables are found in the Results folder (**MoranI_Nsp_nor .csv** for Moran's I under normality and **MoranI_Nsp_ran.csv** for Moran's I under randomization)

References

An, L., Tsou, M.-H., Spitzberg, B., Gawron, J. M., & Gupta, D. K. (2016). Latent trajectory models for space-time analysis: An application in deciphering spatial panel data. *Geographical Analysis*, *48*(3), 314–336. https://doi.org/doi: 10.1111/gean.12097

Chun, Y. (2008). Modeling network autocorrelation within migration flows by eigenvector spatial filtering. *Journal of Geographical Systems*, *10*(4), 317–344. https://doi.org/10 .1007/s10109-008-0068-2

Chun, Y., & Griffith, D. A. (2011). Modeling network autocorrelation in space–time migration flow data: An eigenvector spatial filtering approach. *Annals of the Association of American Geographers*, *101*(3), 523–536.

Chun, Y., Griffith, D. A., Lee, M., & Sinha, P. (2016). Eigenvector selection with stepwise regression techniques to construct eigenvector spatial filters. *Journal of Geographical Systems*, *18*(1), 67–85. https://doi.org/10.1007/s10109-015-0225-3

Claassen, R., Cattaneo, A., & Johansson, R. (2008). Cost-effective design of agri-environmental payment programs: U.S. experience in theory and practice. *Ecological Economics*, *65*(4), 737–752.

Federal Register. (2019). *Conservation reserve program: A rule by the commodity credit corporation.* https://www.federalregister.gov/documents/2019/12/06/2019-26268/ conservation-reserve-program

Gray, R. L., & Teels, B. M. (2006). Wildlife and fish conservation through the Farm Bill. *Wildlife Society Bulletin*, *34*(4), 906–913. https://doi.org/10.2193/0091 -7648(2006)34[906:WAFCTT]2.0.CO;2

Griffith, D. A. (2000). A linear regression solution to the spatial autocorrelation problem. *Journal of Geographical Systems*, *2*(2), 141–156. https://doi.org/10.1007/PL00011451

Hughes, J., & Haran, M. (2013). Dimension reduction and alleviation of confounding for spatial generalized linear mixed models. *Journal of the Royal Statistical Society: Series B (Statistical Methodology)*, *75*(1), 139–159. https://doi.org/10.1111/j.1467-9868.2012 .01041.x

Jonas, H. D., Barbuto, V., Jonas, H.C., Kothari, A., & Nelson, F. (2014). New steps of change: Looking beyond protected areas to consider other effective area-based conservation measures. *PARKS*, *20*(2), 111–128. https://doi.org/10.2305/IUCN.CH .2014.PARKS-20-2.HDJ.en

Lambert, D. M., Sullivan, P., Claassen, R., & Foreman, L. (2007). Profiles of US farm households adopting conservation-compatible practices. *Land Use Policy*, *24*(1), 72– 88. https://doi.org/10.1016/j.landusepol.2005.12.002

Maxwell, S. L., Cazalis, V., Dudley, N., Hoffmann, M., Rodrigues, A. S. L., Stolton, S., Visconti, P., Woodley, S., Kingston, N., Lewis, E., Maron, M., Strassburg, B. B. N., Wenger, A., Jonas, H. D., Venter, O., & Watson, J. E. M. (2020). Area-based conservation in the twenty-first century. *Nature*, *586*(7828), 217–227. https://doi.org /10.1038/s41586-020-2773-z

Mishra, A. K., & Khanal, A. R. (2013). Is participation in agri-environmental programs affected by liquidity and solvency? *Land Use Policy*, *35*, 163–170. https://doi.org/10 .1016/j.landusepol.2013.05.015

Ogg, C., & Keith, G. (2002). New federal support for priority watershed management needs. *Journal of the American Water Resources Association, 38*(2), 577–586. https://doi.org/10.1111/j.1752-1688.2002.tb04339.x

Pace, R. K., Lesage, J. P., & Zhu, S. (2013). Interpretation and computation of estimates from regression models using spatial filtering. *Spatial Economic Analysis, 8*(3), 352–369. https://doi.org/10.1080/17421772.2013.807355

Richardson, C. W., Bucks, D. A., & Sadler, E. J. (2008). The conservation effects assessment project benchmark watersheds: Synthesis of preliminary findings. *Journal of Soil and Water Conservation, 63*(6, SI), 590–604. https://doi.org/10.2489/jswc.63.6.590

Riley, T. (2004). Private-land habitat opportunities for prairie grouse through federal conservation programs. *Wildlife Society Bulletin, 32*(1), 83–91. https://doi.org/10.2193/0091-7648(2004)32[83:PHOFPG]2.0.CO;2

Rossi, G. D., Hecht, J. S., & Zia, A. (2021). A mixed-methods analysis for improving farmer participation in agri-environmental payments for ecosystem services in Vermont, USA. *Ecosystem Services, 47*, 101223. https://doi.org/10.1016/j.ecoser.2020.101223

Tobler, W. R. (1970). A computer movie simulating urban growth in the Detroit region. *Economic Geography, 46*(2), 234–240.

USDA Farm Production and Conservation Business Center. (2020). *County-level CRP and EQIP dataset in the USA*. Economics and Policy Analysis Division, Data Services Branch.

USDA Farm Service Agency. (2019). *News release of USDA to Open Signup for Conservation Reserve Program on December 9 [Government news release]*. https://www.fsa.usda.gov/news-room/news-releases/2019/usda-to-open-signup-for-conservation-reserve-program-on-december-9

Wilson, G. V., Shields, F. D., Bingner, R. L., Reid-Rhoades, P., DiCarlo, D. A., & Dabney, S. M. (2008). Conservation practices and gully erosion contributions in the Topashaw Canal watershed. *Journal of Soil and Water Conservation, 63*(6), 420–429. https://doi.org/10.2489/jswc.63.6.420

4   Concurrent green initiatives in China

China is the most populous country, with its territory area the third largest in the world. Its economic boom over the past four decades has led to its gross domestic product (GDP) reaching US$17.7 trillion in 2021, making China's economy the second largest on the earth (China Briefing, 2022). On the other hand, China's environmental problems are among the most severe of any major countries and will likely worsen. The Chinese people, including the top leaders, are aware of these challenges and have spent a large amount of efforts and time handling such problems. The positive consequences are observable—for instance, the air quality in Beijing and other major cities—has become better. However, such efforts are not great enough to counter the forces that are driving environmental degradation and destruction in the country. The evidence comes from the deterioration of many essential indicators, which include—but are not limited to—biodiversity losses, depleted fisheries, grassland degradation, cropland losses, rapid desertification, disappearing wetlands, increasing frequency and magnitude of human-induced natural disasters, soil erosion, interrupted river flow, salinization, and water pollution and shortages (Liu & Diamond, 2005; Xu et al., 2019a, b).

China launched its Grain-to-Green Program (GTGP) in 1999–2001 and the Forest Ecological Benefit Compensation (FEBC) Fund in 2001. Since 2004, these two programs have been implemented simultaneously in 20 provinces, autonomous regions, and municipalities. GTGP-eligible land parcels are farmland on steep slopes, whereas FEBC parcels are natural forestlands, thus spatially disconnected from GTGP parcels. In many regions, parcels of both types of land are contracted to the same households (Yost et al., 2020), making them horizontally stacked payments.

4.1 Grain-to-Green Program (GTGP)

The Chinese central government proposed the Grain-to-Green Program (Phase I) in 1999 in some parts of China. In the upper reach of the Yangtze River Basin and the upper and middle reaches of the Yellow River Basin, the government paid farmers 2,250 and 1,500 kg of grain (around 3,150 and 2,100 yuan, respectively,

DOI: 10.4324/9781003290292-4

at a price of 1.4 yuan per kg of grain) per year for each hectare of converted cropland. Farmers may then receive additional funding, including 300 yuan/ha per year for miscellaneous expenses and a one-time payment of 750 yuan/ha for seeds or seedlings. Because the targeted croplands are primarily on steep slopes, GTGP is also known as the Sloping Land Conversion Program (SLCP) in literature. China's National Forestry and Grassland Administration refers to this program as the Conversion of Cropland to Forest Program, which covers all croplands enrolled in the program. The duration of subsidies varies depending on cropland conversion: two years if converted to grassland, five years if converted to economic forests by using fruit trees, or eight years if converted to ecological forests by using tree species such as Chinese pine (*Pinus tabuliformis*) and black locust (*Robinia pseudoacacia*). Furthermore, no taxes on the converted cropland were collected.

With the eight-year extension of the GTGP in 2007 (Phase II), the compensation was reduced by nearly half: 2025 yuan/ha per year in south China and 1,425 yuan/ha per year in north China with no grain subsidy available. The actual amount of compensation varied from province to province. For example, Guizhou province's compensation was cash, and grain subsidies were replaced by money at equivalent market value. The total compensation was 3,585 yuan/ha for the first eight years (Phase I) and 2,010 yuan/ha for the eight-year extension (Phase II). The amount of compensation received by farmers at different places may vary because local village leaders might divert varying proportions of the money to other purposes.

GTGP has produced enormous ecological and socioeconomic benefits at local, regional, national, and global scales (Liu et al., 2008; Zhang et al., 2010). In rural areas, poor households need financial support, even small in amount, to afford the costs (e.g., transportation) associated with migration. This support explains the positive relationship between household income and migration propensity. However, when household income goes above a certain level, the link to migration may become weak and even reverse: the higher the income, the less likely the relevant household may migrate out. Observed in many parts of the world, the "inverted U-shape relationship" describes how income relates to the propensity for migration (Dao et al., 2018; Zhao, 2003).

4.2 Forest Ecological Benefit Compensation (FEBC)

The Chinese central government launched the experimental phase of its Forest Ecological Benefit Compensation (FEBC) program in 2001 in 11 provinces, autonomous regions, and municipalities, covering around 200 million ha (Deng et al., 2011; Ouyang et al., 2013). The FEBC program started formally in 2004, according to the Forestry Law of the People's Republic of China and the Decision to Promote Forestry Development by the Central Committee of the Chinese Communist Party and State Council. The program aims to establish, nurture, protect, and manage ecological welfare forests (EWF; Dai et al. 2009), i.e., forestland with vulnerable yet essential ecological benefits (Dai et al., 2009; Ministry of Finance & State Administration of Forestry, 2007). The FEBC program, the Natural Forest Conservation Program (NFCP), Grain-to-Green Program (GTGP), and Ecological Transfer Payment (ETP) are significant components of China's forest eco-compensation mechanism (Ouyang et al., 2013).

Two components comprise the FEBC program: the national ecological welfare forest (EWF) fund from the central government and the local ecological welfare forest fund from the local governments (Deng et al., 2011). By definition, the national EWF fund covers important nationwide forestlands approved by the former State Forestry Administration, while the local EWF fund covers regionally important forestlands identified and approved by provincial or same-level governments. By the end of 2006, 25 provinces, autonomous regions, or municipalities had set up various local EWF funds (Deng et al., 2011). However, this article only considers national EWF land as "few provincial governments have committed subsidies to protect local EWF lands" (Dai et al., 2009). By the end of 2006, FEBC had protected a total of 104 million ha (1,560 million mu) of the national EWF forestland. However, according to Ouyang et al. (2013), the above area was 70 million ha, accounting for an accumulative investment of over 20 billion yuan by 2013.

The initial compensation was 5 yuan/mu (75 yuan/ha) for national EWF forestland, 4.75 yuan/mu used for protection and management by the corresponding forestry entrepreneur, community, or individuals, and 0.25 yuan/mu for governmental expenses, fire protection and road maintenance. Starting from 2010, the Chinese central government increased the compensation standard with a differential compensation policy: For national EWF forestland owned (note: partial ownership; essentially all land is owned by the central government) by collective organizations or individuals, the compensation was 9.75 yuan/mu and 0.25 yuan/mu for governmental expenses, fire protection, and road maintenance (Ministry of Finance & State Administration of Forestry, 2010).

4.3 Comparison between GTGP and FEBC programs

Below we compare China's Grain-to-Green Program (GTGP) and Forest Ecological Benefit Compensation (FEBC) programs (Table 4.1). Aside from the differences identified in the table, FEBC participation is more heavily prescribed than GTGP participation (Yost et al., 2020). Given that both GTGP and FEBC consider the protection of soil erosion as a paramount goal and share similar land eligibility standards (Dai et al., 2009), they occur concurrently in most places in China, which is evidenced by our Fanjingshan (Chapter 5) and Tianma cases (Chapter 6).

4.4 Area-based conservation experiment

If there is no spillover effect, Figure 4.1A shows the green initiatives in three types of areas: GTGP-eligible only areas, GTGP and FEBC simultaneously

Table 4.1 Comparison of China's GTGP and FEBC programs

Program	GTGP	FEBC (for national ecological welfare forest or EWF)
Program goal	Restore vegetation and reduce ground runoff and soil erosion	Protect existing forests and seek ecological security by reducing water runoff and erosion
Qualification	Cropland with slope ≥15° in northwestern China and ≥25° elsewhere	Slope ≥16° in Northeast China and ≥26° in South-central and other areas of China; also consider vegetation and precipitation (Dai et al., 2009)
Start year	1999 (variable by location)	2001 (variable by location)
Compensation (yuan/ mu/year)	239[a]	14.75[b]
Obligations	Convert cropland to forestland or grassland	Prohibit forest fire, illegal logging, and poaching
Average cropland enrolled per household[c] (mu)	3.89	57.25

Notes: The table is modified from table 1 of Yost et al. (2020).

[a] This payment rate dropped to 134 yuan/mu for Phase II of the program from 2007 to 2015. It is also subject to change depending on year and place.

[b] This rate applies to forestland contracted to individual households. A different rate (5 yuan/mu) applies to state-owned forestland. It is also subject to change depending on year and place.

[c] These numbers are from 200 households (out of our sample of 605) that enrolled land in both GTGP and FEBC at Fanjingshan National Nature Reserve based on a survey conducted in 2014 (Yost et al., 2020).

eligible areas, and FEBC eligible only areas. Due to the spillover effects, an extra amount of GTGP enrollment in GTGP and FEBC simultaneously eligible areas may come out, as discussed earlier, the oval in Figure 4.1B.

These findings may be good news: if other conditions are met, we should invest more green efforts in the middle area. Our data show that some local households may be free riders: they enroll more land in GTGP for more money but fail to fulfill their obligations. So we suggest policymakers change the enrollment rules such that the FEBC efforts can be reallocated to FEBC-only areas, i.e., the dotted FEBC effort in Figure 4.1C. As a result, the extra GTGP enrollment is gone, avoiding payments with zero or little ecological benefits. The other benefit is through a reshuffle of green efforts, we can allow a 3% budget cut for the whole country but still keep the total area of GTGP and FEBC unchanged. These considerations may be vital during times of crisis, such as COVID-19.

Next, we performed a green initiatives-reallocation experiment as we did for EQIP and CRP in Chapter 3: in a standard year (i.e., with relative stable GTGP

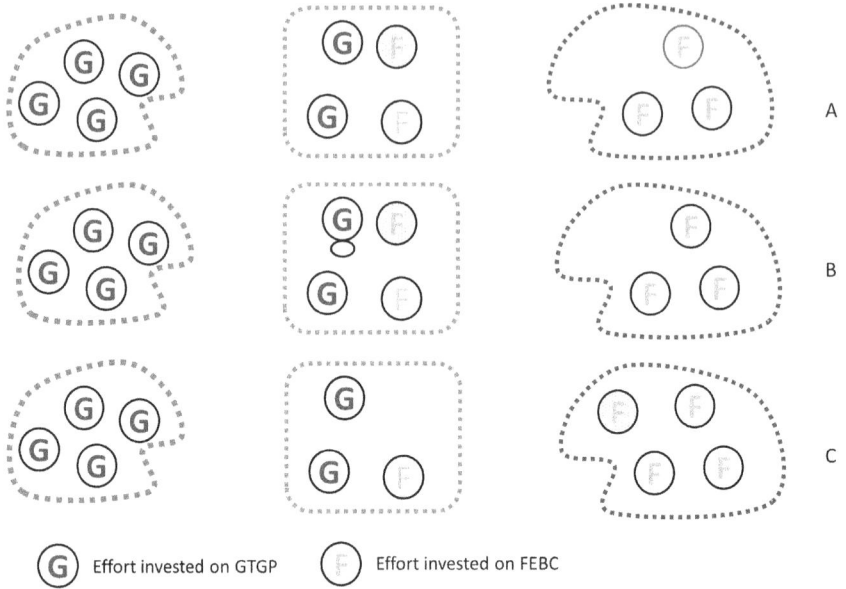

Figure 4.1 Initiative efforts invested in GTGP and FEBC under three conditions: (A) there are no spillover effects, (B) there is a spillover effect, and (C) the spillover generating FEBC effort is relocated to FEBC-only areas. The left, middle, and right dotted areas represent GTGP-eligible only areas, GTGP and FEBC simultaneously eligible areas, and FEBC eligible only areas.

enrollment), there is 5.51 million ha of GTGP land, which can decompose to 0.47 (due to FEBC) and 5.04 (original GTGP by its own) million ha (Section 5.4). For this 0.47 million ha GTGP land that comes as a consequence of nearby FEBC land, we can "reallocate" its 25%, 50%, 75%, and 100% back to FEBC. As a result, the total area of GTGP and FEBC does not change (i.e., it keeps at 5.51 million ha), but the total payment decreases with an increasing amount in FEBC. The much lower pay rate of FEBC may account for this decline in total combined payments of GTGP and FEBC. The results show that we can allow a budget cut of 0.79–3.17% without affecting the total amount of land devoted to GTGP and FEBC (Table 4.2).

Table 4.2 Payments saved due to policy redesign between GTGP and FEBC

| | Pre-pandemic | | During or post-pandemic | | | | | | | |
| | Baseline[a] | | 25% relocation[b] | | 50% relocation | | 75% relocation | | 100% relocation | |
	Area	Pay	Area	Pay	Area	Pay	Area	Pay	Area	Pay
GTGP[c]	5.51	1,683.14	5.39	1,647.25	5.28	1,611.35	5.16	1,575.46	5.04	1,539.57
FEBC[c]	70	2,353.72	70.12	2,357.67	70.24	2,361.62	70.35	2,365.57	70.47	2,369.53
Total	75.51	4,036.86	75.51	4,004.92	75.51	3,972.98	75.51	3,941.04	75.51	3,909.09
Change	0	/	0	−31.94	0.00	−63.88	0.00	−95.83	0.00	−127.77
Change %	0	/	0	−0.79	0.00	−1.58	0.00	−2.37	0.00	−3.17

Notes:
[a] We use the FEBC data in 2013 and GTGP data in 2006 (area: million ha; pay: million $) due to data availability.
[b] FEBC has generated 0.47 million ha of GTGP land.
[c] The average pay rate for GTGP and FEBC is 134 Yuan/mu (305.47 $/ha) and 14.75 Yuan/Mu (33.6246 $/ha), respectively.

References

China Briefing. (2022). *China's most productive provinces and cities as per 2021 GDP statistics*. https://www.china-briefing.com/news/chinas-2021-gdp-performance-a-look-at-major-provinces-and-cities/

Dai, L., Zhao, F., Shao, G., Zhou, L., & Tang, L. (2009). China's classification-based forest management: Procedures, problems, and prospects. *Environmental Management, 43*(6), 1162–1173.

Dao, T. H., Docquier, F., Parsons, C., & Peri, G. (2018). Migration and development: Dissecting the anatomy of the mobility transition. *Journal of Development Economics, 132*, 88–101.

Deng, H., Zheng, P., Liu, T., & Liu, X. (2011). Forest ecosystem services and eco-compensation mechanisms in China. *Environmental Management, 48*(6), 1079–1085.

Liu, J., & Diamond, J. (2005). China's environment in a globalizing world. *Nature, 435*(7046), 1179–1186.

Liu, J., Li, S., Ouyang, Z., Tam, C., & Chen, X. (2008). Ecological and socioeconomic effects of China's policies for ecosystem services. *Proceedings of the National Academy of Sciences, 105*(28), 9477–9482. https://doi.org/10.1073/pnas.0706436105

Ministry of Finance, & State Administration of Forestry. (2007). *Central Government sponsored forest ecological benefit compensation fund: Management rules*. http://www.mof.gov.cn/zhengwuxinxi/caizhengwengao/caizhengbuwengao2007/caizhengbuwengao20075/200805/t20080519_26401.html

Ministry of Finance, & State Administration of Forestry. (2010). *Central Government sponsored forest ecological benefit compensation fund: Management rules* (Revised ed.). http://extwprlegs1.fao.org/docs/pdf/chn152418.pdf

Ouyang, Z., Zheng, H., & Yue, P. (2013). Establishment of ecological compensation mechanisms in China: Perspectives and strategies. *Acta Ecologica Sinica, 33*(3), 686–692.

Xu, W., Fan, X., Ma, J., Pimm, S. L., Kong, L., Zeng, Y., Li, X., Xiao, Y., Zheng, H., Liu, J., Wu, B., An, L., Zhang, L., Wang, X., & Ouyang, Z. (2019a). Hidden loss of Wetlands in China. *Current Biology, 29*(18), 3065-3071.e2. https://doi.org/10.1016/j.cub.2019.07.053

Xu, W., Pimm, S. L., Du, A., Su, Y., Fan, X., An, L., Liu, J., & Ouyang, Z. (2019b). Transforming protected area management in China. *Trends in Ecology & Evolution, 34*(9), 762–766. https://doi.org/10.1016/j.tree.2019.05.009

Yost, A., An, L., Bilsborrow, R., Shi, L., Chen, X., & Zhang, W. (2020). Linking concurrent payments for ecosystem services in a Chinese nature reserve. *Ecological Economics, 169*, 106509.

Zhang, Q., Bennett, M. T., Kannan, K., & Jin, L. (2010). Payments for ecological services and eco-compensation: Practice and innovations in the People's Republic of China. In Proceedings from the International Conference on Payments for Ecological Services. International Conference on Payments for Ecological Services, September 6–7, 2009, Mandaluyong City, Philippines.

Zhao, Y. (2003). The role of migrant networks in labor migration: The case of China. *Contemporary Economic Policy, 21*(4), 500–511.

5 Concurrent green initiatives in Fanjingshan National Nature Reserve, China

China's two large green initiatives, Grain-to-Green Program (GTGP) and Forest Ecological Benefit Compensation (FEBC) program, overlap in space in many provinces. Potential spillover effects might occur. This chapter seeks to answer whether spillover effects may exist and, if so, what are the significant characteristics of such effects and the underlying mechanisms.

5.1 Fanjingshan National Nature Reserve

We used the Fanjingshan National Nature Reserve (N27°44'-28°03', W108°34'-108°48'), China (Figure 5.1), as our first study site to explore potential spillover effects between GTGP and FEBC. The reserve is in Guizhou province in southwestern China, with around 419 km². The reserve was established in 1978 as a protected area for the Guizhou golden monkey (*Rhinopithecus brelichi*) and then extended to its current size in 1986 to conserve other animal and plant species within the reserve. According to China's Wild Animal Protection Law, many of these species are listed as first-class or second-class protected wildlife species; exemplar species include the clouded leopard (*Neofelis nebulosa*) and the Asiatic black bear (*Ursus thibetanus*).

As a flagship reserve of relatively undisturbed subtropical ecosystems constituting part of the 25 global biodiversity hotspots (GDF & FNNR, 1990), the reserve is replete with over 6,000 plant, animal, and bird species (GEF Project Team, 2004). Established in 1978 and then extended to its current size in 1986, it is also home to the last and only population (around 750 animals) of the Guizhou snub-nosed monkey (also named Guizhou golden monkey; *Rhinopithecus brelichi*), an umbrella, endangered species susceptible to human presence, activity, and the resultant habitat degradation (Yang et al., 2002).

Fanjingshan has over 13,000 people who live a subsistence lifestyle, growing crops and vegetables and raising pigs and other livestock. Around 70% of these people belong to ethnic minorities, such as Tujia and Miao (GEF Project Team, 2004, p. 8). Depending on many local natural resources such as Chinese medicine herbs, fuelwood, timber, and bamboo, local people can enter non-core habitat areas to collect these resources and herd livestock. Illegal wood extraction and poaching are occasionally reported (An et al., 2020).

DOI: 10.4324/9781003290292-5

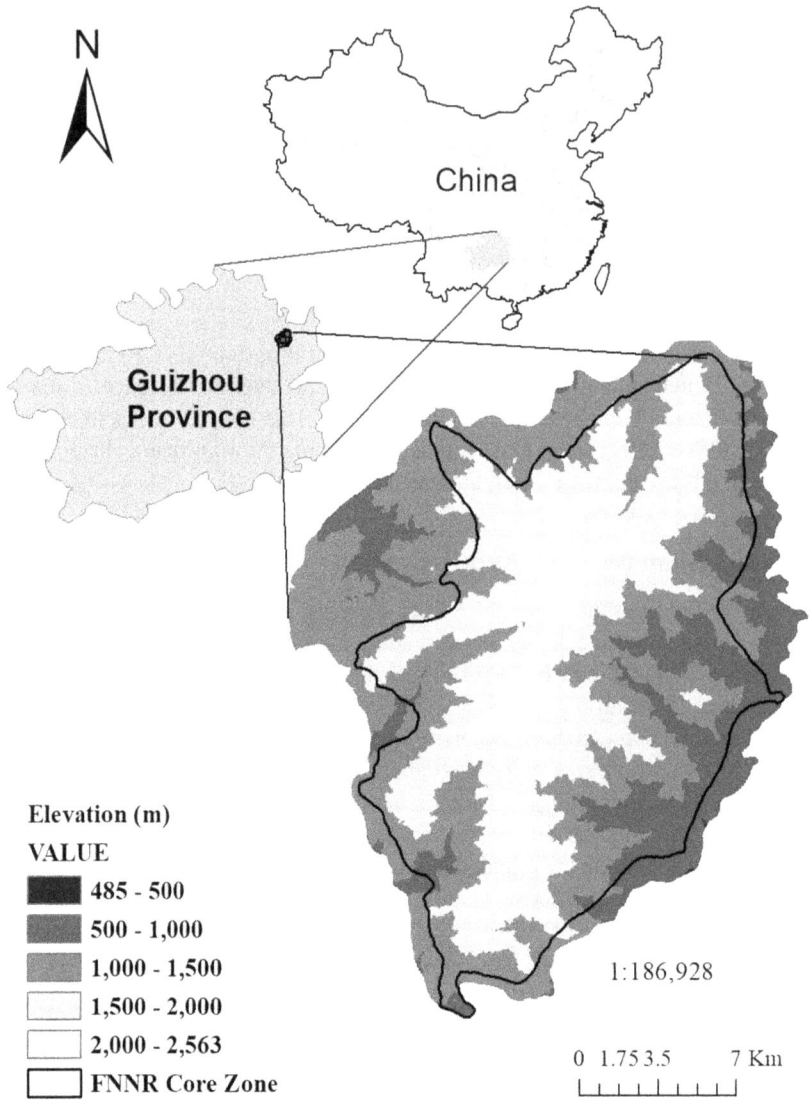

Figure 5.1 Location of Fanjingshan National Nature Reserve, China.

Like many other rural areas in China, the economic boom has led to drastic land-use changes, including deforestation and reforestation. The last decade has witnessed rapid tourism development and demographic transition—many young people migrate to cities for higher-pay jobs and send remittances to their family members left behind. Fanjingshan is a locale that implements the Forest Ecological Benefit Compensation (FEBC) and the Grain-to-Green Program

(GTGP). Fanjingshan initiated the GTGP program around 2000, at which time 774 households, primarily inside the reserve, enrolled land parcels in the program. In Fanjingshan, farmers planted pine, Chinese fir, bamboo, and other species with the local government's seedlings. Fanjingshan started FEBC in 2002, using government funding for local people's patrol of designated forests (there was no forest enterprise at Fanjingshan due to its nature reserve status). FEBC money was either paid to individual villagers or allocated to local communities such as administrative groups. Local people, including Fanjingshan officials, primarily hold a favorable view toward FEBC and GTGP, crediting them to improve the local environment and promote wildlife conservation.

5.2 Data collection

We used a stratified random sampling strategy to select households for interviews in 2014 and 2015 (An et al., 2020; Yost et al., 2020). Our sampling frame was a roster of all 3,256 households, which was based on the census of Fanjingshan in 2013. An administrative village is comprised of several natural villages (also named villager groups or resident groups). A villager group is mainly equivalent to a "production team" made up of closely knit households in charge of cultivating collective cropland before implementing the Household Registration System in the early 1980s.

We subdivided the whole 3,256 households into 123 sampling units, out of which we randomly selected 58 units. We randomly assigned these 58 sampling units to 20 administrative villages within or across Fanjingshan boundaries. The sampling units' assignment was in proportion to each administrative village's population size, while smaller administrative villages were slightly over-represented in our sample. Then at each administrative village, we took a random sample according to the number of sampling units it received on a 20 households per unit basis: if an administrative village received an assignment of one unit, we then randomly selected 20 households for an interview; if two units, then 40 households, and so on.

This sampling strategy ended up with 1,160 households pre-selected as our sampling pool, more extensive than our intended sample size of 650. This choice was due to various practical challenges such as the inability to find a knowledgeable person in a household, the absence of household members due to travel or outmigration, or unwillingness to participate in our interview. We eventually managed to complete the entire survey of 605 households in 2014. We skipped the content of our 2014 survey as our paper does not use its data.

Since some households remained unreachable (e.g., no knowledgeable member available at the time of our visit), we ended up with a survey of 494 households in 2015. The survey questionnaire consisted of three sections: (1) household land-use; (2) participation in GTGP, FEBC, and other programs, if any; and (3) household time allocation, amount of harvest, and cash income (if sold), and gross cash incomes from various activities. Specifically, we inquired about detailed land use, including the total area of dryland and paddyland, the portion of land enrolled at GTGP, walking distance from home to each plot, and the productivity of these plots before or after GTGP.

We also mapped all relatively big (greater than 0.1 mu) GTGP and non-GTGP plots on Google Earth and converted them into regular GIS shapefiles. We chose three plots from GTGP land, representing the farthest, medium, and closest plot to the household, and similarly chose three more plots from non-GTGP land in our sample. Then for each selected plot, we asked their willingness to enroll or reenroll (if already in GTGP) under a set of hypothetical conditions regarding GTGP payment level, duration, post-enroll land-use options, and neighbors' attitudes toward GTGP (Yost et al., 2020).

In the 2015 survey, we inquired about detailed information on their reserved and responsible forestland, called "self-maintained and responsibility mountains", respectively, that may receive FEBC payments. The former refers to the small parcel(s) of forestland assigned to the household before China's rural reform in the early 1980s (Krusekopf, 2002), while the latter refers to the forestland (usually much more extensive in size) allocated to the corresponding household after the rural reform. We collected detailed data about the area, geographic location (via Google Earth), resource collection, amount of funds received, and patrol efforts (frequency, time, etc.). We also collected information about participants' willingness to patrol under a set of hypothetical policy conditions (i.e., patrol frequency, compensation level, and neighbor's willingness to patrol). For more information about the variables, see Table 5.1.

5.3 Modeling the amount of land enrolled in GTGP

We modeled the total area of land enrolled in the GTGP by a household (GP_TL_Amt) as a function of the variables listed in Table 5.1 (not including plot level data) using OLS regression. The independent variables (Table 5.1) in our model were selected based on the Sustainable Livelihoods Framework, in which human, social, natural, physical, and financial capitals that a household possesses play a crucially important role in relevant livelihood decisions (United Nations Development Programme, 2017). We did not collect data for physical capital (such as access to road, clean water, information, and affordable energy) as they do not vary substantially in our study area.

At Fanjingshan, GTGP was implemented earlier than FEBC. In addition, local farmers had more freedom to participate in GTGP or decline it, while FEBC participation was largely government-prescribed. We thus choose GTGP enrollment land as the dependent variable and FEBC land as an independent variable with control of several other variables. The regression takes the following form:

$$y = b_0 + b_1 X_1 + \sum_{j=1}^{k} c_j CV_j + e \tag{5.1}$$

where y is the area of cropland enrolled in GTGP, X_1 is the area of land enrolled in FEBC, CV_j are the controlled variables that represent influences from a set of

Table 5.1 Definition and descriptive statistics of variables at Fanjingshan, China

Variable name	Definition (unit)	Min	Max	Mean
GP_TL_Amt	Area of land the household enrolled in GTGP (mu)	0	15	2.38
lgGPTLLd	Logarithm of GP_TL_Amt[a]	−13.85	2.71	−3.28
FstMnyAmt	Annual FEBC payment a household receives (1,000 yuan)	0	3.67	0.58
lgFstMnyAmt	Logarithm of FstMnyAmt	−13.82	1.30	−4.99
DryLdAmt	Amount of dryland the household has (mu)	0	22.5	2.67
LgHDryLd	Logarithm of DryLdAmt	−13.82	3.11	0.38
PadLdAmt	Amount of paddyland the household has (mu)	0	20	3.47
LgHPadLd	Logarithm of PadLdAmt	−13.82	3.00	0.06
HhCshInc	Household cash income in 2014 (1,000 yuan)	0	1230.97	60.59
LgHCshInc	Logarithm of HhCshInc	−13.82	7.12	2.91
AllFstAmt	Amount of FEBC forestland the household has (mu)	0	1215.00	18.36
LgFstAmt	Logarithm of AllFstAmt	−13.82	7.10	2.91
HH_Size	Household size (# of people in the household)	1	9	3.20
HHLbr	Household labor (# of people with age from 15 to 59 years)	0	6	2.21
PlotInGP	Plot already in GTGP (1 for yes and 0 for no)	0	1	0.40
Plot_Dst	Distance from plot to the household (minutes of walking)	0.08	270.00	23.24
Plot_Area	Area of plot (mu)	0.05	15.00	0.77
Plot_Mny	Hypothetical amount of GTGP pay (100 yuan)	1	7	4
Plot_Span	Hypothetical amount of GTGP span (years)	4	12	7.99
Fallow	Land parcel left fallow (0 for not and 1 for fallow)	0	1	0.25
NB_Pct	Percent of neighbors agreed to join GTGP (percent)	25	75	49.15

Note:
[a] Some households have extremely large numbers. Taking a logarithm makes the associated variables less skewed and more appropriate for later regression analysis. All the variables with the name beginning with Lg are such logarithm-transformed variables.

carefully chosen household-level capitals, c_j the coefficients of these variables, and e the residual.

Human capital is represented by household size (HH_Size), household labor (HHLbr), and the amount of land allocated to the corresponding household during the time of the household responsibility contract system around 1980 (Krusekopf, 2002). The area of dryland (DryLdAmt) and area of paddyland (PadLdAmt) represent natural capital. We differentiated dryland and paddyland as the former may be farther away from households, in sloping areas, and more likely subject to GTGP enrollment. These two areal land measurements are essential as they determine—at least affect—available land supply and food security for a household (Joshi, 2011; Yost et al., 2020).

For social and/or financial capitals, we first considered concurrent PES policy, which was found to affect households' behavior (Yost et al., 2020). Since local villagers were not involved in FEBC decisions (the local government essentially determined who may enroll and how much land should go into enrollment), we only examined whether and how the amount of FEBC payments (FstMnyAmt) or area of forestland enrolled in FEBC (AllFstAmt) may affect GTGP enrollment (GP_TL_Amt) when controlling for a set of relevant variables. We selected the FEBC compensation amount (FstMnyAmt) and the amount of enrolled land (AllFstAmt). Note that the FEBC payment amount (FstMnyAmt) and land amount (AllFstAmt) were not highly correlated ($r=0.0972$, $p=0.10$); therefore, we included both in the regression without much concern about multicollinearity. Second, we used an income variable, i.e., household cash income (HHCshInc), to represent the impact of financial capital. To reduce the potential adverse effects of a skewed distribution (Lo & Andrews, 2015; Olivier et al., 2008), we also calculated the logarithm of these variables (Table 5.1).

Following relevant literature (Chen et al., 2009a, b) and our work earlier (Yost et al., 2020), we included several variables that represent plot-level characteristics, including plot enrollment in GTGP before 2015 (PlotInGP), distance from plot to the household (minutes of walking; Plot_Dst), area of the plot (mu; Plot_Area), hypothetical amount of GTGP pay (100 yuan; Plot_Mny), hypothetical period of GTGP enrollment (years; Plot_Span), and whether the plot would be left fallow if it were enrolled in GTGP (0 for not and 1 for fallow; Fallow). To explore the impacts of social norms, we included a variable, the percent of neighbors who would agree to join GTGP (NB_Pct), following relevant literature (Chen et al., 2009a, b) and our previous work (Yost et al., 2020). Note that these plot-level variables (i.e., from PlotInGP to NB_Pct; Table 5.1) are only applicable to our modeling of the willingness to enroll land in GTGP under a set of hypothetical policy variables.

Examining the regression results (Table 5.2), we found that FEBC payment (FstMnyAmt), a variable representing concurrent payments for environmental services, exerted a significant impact on the area of land enrolled in GTGP (coefficient$=0.4487$, $p=0.0644$, significant at $\alpha=0.10$ level). For reasons behind this relationship, we refer to Yost et al. (2020) and Section 6.4 where we found similar results at Tianma National Nature Reserve (Chapter 6).

5.3.1 Modeling the logarithm of the amount of land enrolled in GTGP

Many stark differences between households made the data difficult to analyze. These households were extremely rich or poor, had a very big or small area of dryland, paddyland, or FEBC forestland, or received very high or low FEBC payments, making related data very skewed (see Table 5.1). We, therefore, analyzed

Table 5.2 Regression results for GTGP at Fanjingshan, China

Variable	Description	Coefficient	p-value	VIF
Intercept	—	0.4469	0.2686	0
DryLdAmt	Dryland amount	0.5839	<0.0001	1.0830
PadLdAmt	Paddyland amount	0.2737	<0.0001	1.0438
HHCshInc	Household cash income	−0.0017	0.2692	1.0423
FstMnyAmt	FEBC payment amount	0.4487	0.0644	1.1026
HH_Size	Household size	0.1530	0.2710	2.3787
HHLbr	Number of laborers (age between 15 and 59)	−0.1295	0.5461	2.3700
AllFstAmt	FEBC forestland amount	−0.0014	0.4675	1.0751
R^2 (Adjusted R^2)	0.3827 (0.3666)			

Note: This model does not have the logarithm change of variables; the dependent variable name is GP_TL_Amt.

Table 5.3 Modeling results for the logarithm of GTGP (lgGPTLLd) at Fanjingshan, China

Variable	Description	Coefficient	p-value	VIF
Intercept		0.5253	0.0009	0
LgHDryLd	Log of dryland area	0.2735	<0.0001	1.0317
LgHPadLd	Log of paddyland area	0.1298	0.0003	1.0269
LgHCshInc	Log of cash income	−0.0440	0.1054	1.0354
LgFstMnyAmt	Log of FEBC payment	0.0162	0.0755	1.0253
HH_Size	Household size	0.0873	0.1008	2.3919
HHLbr	Number of laborers (age between 15 and 59)	−0.0383	0.6405	2.3889
LgFstAmt	Log of FEBC forestland area	0.0121	0.2970	1.0373
R^2 (Adjusted R^2)	0.2123 (0.1918)			

the data after a logarithmic transformation of the independent variables to avoid potential problems arising from skewed data and the disproportional impact of significant outliers on regression results (Lo & Andrews, 2015; Olivier et al., 2008). The results are in Table 5.3.

The impact of FEBC land area did not change after the logarithm transformation: it was still a positive predictor (coefficient=0.0162) and significant at $\alpha=0.10$ level ($p=0.0755$; Table 5.3). Also, interestingly, household size and household labor are both insignificant. We also see decreases in R^2 and adjusted R^2 due to the logarithm transformation. All these changes in regression results confirmed our concern that outliers may change regression results to a considerable extent. Despite such an influence, the impact of the concurrent PES program (log-transformed FEBC payment: coefficient=0.0162, $p=0.0755$; Table 5.3) remained significant, suggesting a similar significant coefficient (coefficient=0.4487, $p=0.0644$; Table 5.2) was not an outcome due to data skewness. Later in Section 6.4, a similar outcome arose from Tianma National Nature Reserve, confirming this spillover effect.

5.3.2 Modeling the impact of FEBC payment on willingness to enroll land in GTGP

We also modeled the willingness to enroll in GTGP under a set of hypothetical conditions using the discrete choice modeling technique. According to Lancaster's approach to consumer theory, attributes determine the utility of certain goods or services rather than the goods or services per se (Lancaster, 1966). According to this theory, the stated preferences (regarding participation in GTGP) from our household interviews can be modeled based on a random utility model (RUM) specification (McFadden, 1974). The RUM is potent for quantifying the preferences of individuals when they make decisions of choosing a particular product or service from a finite set of alternatives. A simple operating assumption exists that the service or product chosen by the consumer yields the highest utility among all alternative services or products available in the corresponding choice set. This model is powerful in our situation because the respondents had to choose one out of two alternative decisions: either participating in the GTGP or not given a set of controlled variables (Yost et al., 2020). Similar modeling efforts can be found elsewhere (Chen et al., 2009, b).

In light of this insight, we collected data about several key GTGP program attributes during our interviews of 605 households in 2014 and 494 households in 2015. The data collected in 2015 and the discrete choice model we developed in this paper were very similar to those in 2014, and we refer to Yost et al. (2020) except for the difference discussed below.

The results based on the survey data in 2015 (Table 5.4) were mainly consistent with those from the 2014 survey (Yost et al., 2020) despite a substantial difference in the questions we asked. In the 2014 survey, a respondent was presented with a hypothetical scenario regarding GTGP policy features. Hypothetical conditions included combinations of a randomly chosen value for payment (Mny), payment span (Plot_Span), post-enrollment land-use choice (fallow, planting cash trees, or planting ecological trees), and neighbors' choice regarding GTGP participation (NB_pct). After being presented with the hypothetical scenario, then the question was formed: "under this combination of hypothetical values, would you be willing to enroll *part of your farmland* in the assumable GTGP?" The question did not target a specific parcel of the land.

In 2015, we selected a set of farmland plots, including those already enrolled in GTGP and ones not enrolled yet at the survey time. Then for each specific plot, we asked the following slightly modified question: "under this combination of hypothetical values, would you be willing to enroll *this specific farmland plot* in the assumable GTGP?"

Under various hypothetical conditions, local villagers still considered the area of available dryland a critical, positive predictor (0.0989; $p = 0.0446$; Table 5.4) for the GTGP enrollment decision, but not paddyland (0.0612, $p = 0.1221$; Table 5.4).

Table 5.4 Modeling results of the impacts of FEBC on GTGP at Fanjingshan, China

Variable	Description	Coefficient	p-value
Intercept		−1.5041	0.0015
DryLdAmt	Dryland amount	0.0989	0.0446
PadLdAmt	Paddy land amount	0.0612	0.1221
HHCshInc	Household cash income	0.0008	0.5219
FstMnyAmt	FEBC payment amount	0.0471	0.8045
HH_Size	Household size	−0.1273	0.4588
HHLbr	Household labor	−0.1058	0.3418
AllFstAmt	FEBC forestland amount	−0.0028	0.0658
PlotInGP	Plot already in GTGP	0.9544	<0.0001
Plot_Dst	Distance from plot to household	0.0115	0.0003
Plot_Area	Area of plot	0.0630	0.5256
Plot_Mny	Hypothetical amount of GTGP pay	0.1827	<0.0001
Plot_Span	Hypothetical amount of GTGP span	0.0479	0.0556
fallow	Land parcel left fallow	−0.2898	0.0921
NB_pct	Percent of neighbors agreed to join GTGP	0.0121	0.0028
2 Res Log pseudo-likelihood		5,781.79	
Gener. chi-square/DF		0.72	

Note: The model is also called discrete choice modeling, where the dependent variable is My_ Choice (1 for yes and 0 for no for whether to enroll the asked parcel in GTGP or not).

This result may come from the minimal paddyland supply and its importance in maintaining food security. Household cash income was still insignificant in the discrete choice model (0.0008, $p=0.5219$; Table 5.4). Household size and labor availability were insignificant ($p=0.4588$, $p=0.3418$, respectively; Table 5.4), as in the case for modeling current GTGP land (Tables 5.2 and 5.3).

At the plot level, our discrete choice model confirmed the positive impact of GTGP payment on GTGP participation: Each 100 yuan/mu of GTGP payment increased the odds of cropland enrollment in GTGP by 20% ($e^{0.1827}=1.20$ or 120%) when other relevant variables were in control (coefficient $=0.1827$, $p < 0.0001$; Table 5.4). Distance from the parcel to the household (Plot_Dst) and whether a plot was already in GTGP (PlotInGP) at the survey time were both positive predictors, suggesting that if a parcel was farther away or already in GTGP, it was more likely to be enrolled in GTGP. The plot area (Plot_Area) was insignificant, which is understandable as the parcel area might not bear significant importance. The remaining plot-level variables, i.e., hypothetical amount of GTGP time span (Plot_Span) and the land plot left fallow (fallow), had coefficients consistent with those in Yost et al. (2020), and we thus skip their discussion. Worthy of mention is the significant coefficient of the percent of neighbors who would agree to join GTGP (NB_pct; coefficient $=0.0121$, $p=0.0028$), suggesting a certain household's decision to participate in GTGP is highly affected by neighbor decisions. This finding confirms the important effects of social norms (Chen et al., 2009, b).

For every additional mu of FEBC land (AllFstAmt), there is a 0.3% (0.0028; Table 5.4) decrease in the odds of enrolling GTGP because the odds ratio is $e^{(-0.0028)} = 0.9972$. We explain the offsetting impacts of FEBC payment by livelihood strategies under land scarcity: when deciding to enroll additional cropland in GTGP in 2015 (i.e., at the time of the survey), their remaining cropland was relatively scarce (the villagers had the opportunity to enroll a portion of their land in 2001 or later years). Payments from FEBC may offer them cash otherwise available through other sources such as outmigration or enrolling more land in GTGP. Furthermore, food security may be another concern when deciding to enroll some of their cropland (Yost et al., 2020).

We can prove that when the original probability (i.e., the probability before a change in a particular variable occurs and leads to the change in odds) is small, the change in probability is very close to the change in the corresponding odds. Therefore, we conclude that each additional mu of FEBC land should lead to a decrease of 0.30% in the probability of their GTGP land enrollment due to each additional mu of FEBC. At Fanjingshan, the median FEBC area is 10 mu (0.67 ha) (in comparison to 1.5 mu or 0.1 ha, median GTGP land area at Fanjingshan), which can generate a $10 \times 0.30\%$ or 3% decrease in the likelihood of enrolling more land.

5.4 Ecological co-benefits of FEBC program

According to Liu et al. (2013), the total accumulative forestland and grassland due to the implementation of GTGP in China reached 8.80 million ha by the end of 2009. As the total area of GTGP land started to level-off in 2005 (Liu et al., 2008), we assume that China has GTGP-induced farmland and grassland at the magnitude of 8.80 million ha, which is the amount in 2006. We only consider returned cropland and grassland because the portion for barren land does not apply in our situation. Next, we aim to find the total area of farmland out of 8.80 million ha.

According to Wu et al. (2019), the increase of grassland in the Loess Plateau (the Plateau's total area is 625,000 km²), China, due to GTGP between 2000 and 2015, was 5,235.38 km². For the same reason, as mentioned above (level-off of GTGP-induced land since 2005), we consider the Loess Plateau has 5,235.38 km² of GTGP-induced grassland, which equals that in 2006.

According to China's National Bureau of Statistics, its total grassland in 2011 was 393 million ha (3,930,000 km²). Due to the stability in grassland between 2006 and 2011, we consider that China had 3,930,000 km² of grassland in 2006. Assuming the countrywide percentage of grassland converted from cropland is the same as that in the Loess Plateau, China should have gained a total of [5,235.38 × 3,930,000/625,000] = 32,920.07 km², or 3.29 million ha grassland in 2006. Therefore, China's total farmland conversions to forestland due to GTGP was 8.80 − 3.29 = 5.51 million ha, or 82.65 million mu (1 ha = 15 mu).

As mentioned in Sections 4.1–4.3 (regarding China's two major PES programs), a total of 104 million ha (1,560 × 10⁶ mu) of FEBC land was protected in 2006. The consequent annual compensation should then be [1,560 × 10⁶ mu ×

9.75 yuan/mu] = 1.5210 × 10^{10} yuan. Note that for collective or individual-owned FEBC forestland, the compensation was 9.75 yuan/mu.

Our results show that each 1,000 yuan of FEBC payment increased the area of land enrolled in GTGP by 0.4487 mu (Table 5.2), indicating a rate of 4.4870 × 10^{-4} mu/yuan. Therefore, the total "additional" area of GTGP land due to FEBC payment is (1.5210 × 10^{10} yuan) × 4.4870 × 10^{-4} mu/yuan = 6.8247 × 10^6 mu, or 6.8247 million mu. Note that in the mid-1980s, China launched the Classification-based Forest Management system, which was a precursor of the later FEBC (Dai et al., 2009). Out of the total 82.65 million mu of farmland due to GTGP as of 2006, 6.8247 million mu came as a co-benefit of FEBC payment, corresponding to 8.25% of total GTGP land.

Next, we look into the case of Tianma (more detail in Chapter 6). Every 100 mu of FEBC land would generate 0.47 mu more GTGP land (Table 5.4), indicating that each mu of FEBC land would lead to an additional 0.0047 mu of GTGP land. The FEBC pay rate was 8.75 yuan/mu in Tianma, then the rate is [1000 × 0.0047/8.75] = 0.54 mu per 1,000 yuan. Given that the total FEBC land was 104 million ha (1,560 × 10^6 mu) in 2006, the total additional GTGP land enrollment caused by FEBC was 1,560 × 10^6 mu × 0.0047 = 7.3320 × 10^6 mu, or 7.3320 million mu. This amount, which came as a co-benefit of FEBC payment, was 7.3320/82.65 = 8.87% out of 82.65 million mu GTGP forestland in 2008. This result is slightly larger than the above amount based on Fanjingshan results. Therefore, the average co-benefit of GTGP enrollment in China due to FEBC payments is approximately 82.65 million mu × (8.87% + 8.25%)/2 = 7.0748 million mu, which is 0.4717 million ha.

Next, we estimate the reduction in carbon sequestration due to the relationship between GTGP and FEBC. According to Feng et al. (2013), the average annual net ecosystem production (NEP) of woodland in the semi-humid forests was 304.40 g C m^{-2}. Our study site Fanjingshan is in a subtropical climate zone with higher carbon biomass. To derive a conservative estimate (e.g., to be used as a lower bound), we still use the rate of 304.40 g C m^{-2}.

As shown earlier, the FEBC payments have induced an additional enrollment of 6.9343 million mu or 0. 4717 million ha of GTGP land by 2010. The increase in carbon sequestration is estimated to be:

0.4717 million ha × 304.4 g C m^{-2} = 0.4717 × 1,000,000 × 10,000 × 304.40 g C m^{-2} = 143.5850 × 10^{10} g C = 1,435,850 million t C = 1,435.850 billion t C (1 t = 10^6 g)

Next, we calculate what may arise under the hypothetical GTGP policy. Following the same rationale, the FEBC-induced reduction in potential GTGP land if implementing GTGP under the hypothetical conditions was 3.0%, as shown in Section 5.3, which translates to 5.51 million ha (China's total GTGP land) × 0.03 = 0.1653 million ha. The corresponding loss in carbon sequestration is:

0.1653 million ha × 304.4 g C m^{-2} = 0.1653 × 1,000,000 × 10,000 × 304.40 g C m^{-2} = 50.3173 × 10^{10} g C = 503,173 million t C = 503.173 billion t (1 t = 10^6 g).

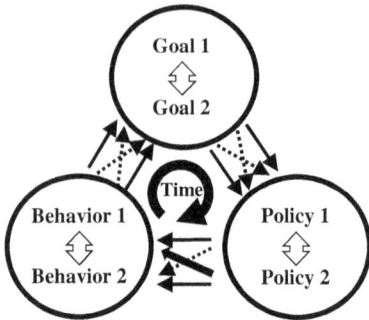

Figure 5.2 Cross-program spillover effects at Fanjingshan, China. The diagram is modified from Figure 1.3, where the solid one-way arrows stand for internal influences from one element to another within the same initiative, while the dashed one-way arrows and double two-way arrows for potential spillover effects; the circular one-way arrow represents Time–Time spillover effects. The big, bold arrow represents the spillover effect with evidence from this section.

5.5 Summary

Fanjingshan National Nature Reserve features a strong *Policy–Behavior* spillover effect, represented by a positive impact of FEBC payment (*Policy* 2) on GTGP enrollment (*Behavior* 1); This effect is represented as the big, bold arrow in Figure 5.2. Extrapolating the average rate to the whole country (for their spatial concurrency in China, see Table 4.1), China should have received 6.93 million mu (0.4623 million ha) "additional" GTGP farmland as an FEBC-induced co-benefit, which could have translated to 1,407,241 million metric tons of carbon sequestration per year under a conservative estimate. However, this positive *Policy–Behavior* spillover effect may turn into a negative one if more land is to be enrolled in GTGP in the future (i.e., under the hypothetical conditions); i.e., the FEBC payments will likely reduce possible enrollment in GTGP by 0.1653 million ha, corresponding to a reduction in carbon sequestration by 503,173 million metric tons.

References

An, L., Mak, J., Yang, S., Lewison, R., Stow, D., Chen, H. L., Xu, W., Shi, L., & Tsai, Y. H. (2020). Cascading impacts of payments for ecosystem services in complex human-environment systems. *Journal of Artificial Societies and Social Simulation, 23*(1), 5.

Chen, X., Lupi, F., He, G., & Liu, J. (2009a). Linking social norms to efficient conservation investment in payments for ecosystem services. *Proceedings of the National Academy of Sciences of the United States of America, 106*(28), 11812–11817.

Chen, X., Lupi, F., He, G., Ouyang, Z., & Liu, J. (2009b). Factors affecting land reconversion plans following a payment for ecosystem service program. *Biological Conservation, 142*(8), 1740–1747.

Dai, L., Zhao, F., Shao, G., Zhou, L., & Tang, L. (2009). China's classification-based forest management: Procedures, problems, and prospects. *Environmental Management*, *43*(6), 1162–1173.

Feng, X., Fu, B., Lu, N., Zeng, Y., & Wu, B. (2013). How ecological restoration alters ecosystem services: An analysis of carbon sequestration in China's Loess Plateau. *Scientific Reports*, *3*, 2846. PMC. https://doi.org/10.1038/srep02846

GDF, & FNNR. (1990). *The study of Fanjingshan (Guizhou Department of Forestry and Fanjingshan National Nature Reserve Administration)*. Guizhou People Press.

GEF Project Team. (2004). *The management plan of Guizhou Fanjingshan National Nature Reserve*. FNNR GEF Project Management Plan Group.

Joshi, L. (2011). A community-based PES scheme for forest preservation and sediment control in Kulekhani, Nepal. In D. Ottaviani and N. E. Scialabba. (Eds.), *Payments for ecosystem services and food security* (pp. 198–203). Rome, Italy: Food and Agriculture Organization of the United Nations.

Krusekopf, C. C. (2002). Diversity in land-tenure arrangements under the household responsibility system in China. *China Economic Review*, *13*(2), 297–312. https://doi.org/10.1016/S1043-951X(02)00071-8

Lancaster, K. J. (1966). A new approach to consumer theory. *Journal of Political Economy*, *74*(2), 132–157.

Liu, J., Li, S., Ouyang, Z., Tam, C., & Chen, X. (2008). Ecological and socioeconomic effects of China's policies for ecosystem services. *Proceedings of the National Academy of Sciences*, *105*(28), 9477–9482. https://doi.org/10.1073/pnas.0706436105

Liu, J., Ouyang, Z., Yang, W., Xu, W., & Li, S. (2013). Evaluation of ecosystem service policies from biophysical and social perspectives: The case of China. In S. A. Levin (Ed.), *Encyclopedia of biodiversity* (pp. 372–384). Academic Press.

Lo, S., & Andrews, S. (2015). To transform or not to transform: Using generalized linear mixed models to analyse reaction time data. *Frontiers in Psychology*, *6*, 1171. https://doi.org/10.3389/fpsyg.2015.01171

McFadden, D. (1974). Conditional logit analysis of qualitative choice behaviour. In P. Zarembka (Ed.), *Frontiers in econometrics*. Academic Press.

National Bureau of Statistics. (2011). *China statistical yearbook 2011*. National Bureau of Statistics.

Olivier, J., Johnson, W. D., & Marshall, G. D. (2008). The logarithmic transformation and the geometric mean in reporting experimental IgE results: What are they and when and why to use them? *Annals of Allergy, Asthma & Immunology*, *100*(4), 333–337. https://doi.org/10.1016/S1081-1206(10)60595-9

United Nations Development Programme. (2017). *Guidance note: Application of the sustainable livelihoods framework in development projects*. United Nations Development Programme Regional Centre for Latin America and the Caribbean. https://www.undp.org/content/dam/rblac/docs/Research%20and%20Publications/Poverty%20Reduction/UNDP_RBLAC_Livelihoods%20Guidance%20Note_EN-210July2017.pdf

Wu, D., Zou, C., Cao, W., Xiao, T., & Gong, G. (2019). Ecosystem services changes between 2000 and 2015 in the Loess Plateau, China: A response to ecological restoration. *PLOS ONE*, *14*(1), 0209483.

Yang, Y. Q., Lei, X. P., & Yang, C. D. (2002). *Fanjingshan research: Ecology of the wild Guizhou snub-nosed monkey (Rhinopithecus bieti)*. Guizhou Science Press.

Yost, A., An, L., Bilsborrow, R., Shi, L., Chen, X., & Zhang, W. (2020). Linking concurrent payments for ecosystem services in a Chinese nature reserve. *Ecological Economics*, *169*, 106509.

6 Concurrent green initiatives in Tianma National Nature Reserve, China

This chapter presents hidden spillover effects between China's Grain-To-Green Program (GTGP) and Forest Ecological Benefit Compensation (FEBC) Fund in Tiantangzhai Township in Anhui Province of China which belongs to a nature reserve called the Tianma National Nature Reserve (TNNR). We draw on household survey data (250 households collected in 2013 and 481 households in 2014) and satellite observations to examine hidden spillover effects among the two concurrent payments for environmental services (PES) programs. We first describe the study site and the two PES programs within the local context, then present the socio-ecological outcomes of the PES programs, and finally summarize the findings and their implications.

6.1 Tianma National Nature Reserve (TNNR)

Tianma National Nature Reserve (TNNR) was set to protect the last remaining patches of secondary natural forests in Southeast China and many protected plant and animal species. The township of Tiantangzhai, which covers an area of 189 km², encompasses the core of the TNNR and spans a geographic extent of N31°05′–31°09′, E115°42′–115°46′ in the eastern Dabieshan Mountain Ranges with elevations ranging from 363 m to 1,729 m (Figure 6.1). Tiantangzhai is located in a subtropical monsoon climate with a mean annual temperature of 16.4°C and mean annual precipitation of 1,350 mm, sustaining lush vegetation. The township is rich with ecotourism resources, which were well developed to attract tourists. Tiantangzhai is home to 4,369 households with a population of 17,295, according to the local household registration record in 2012. The township encompasses 165 resident groups distributed in seven administrative villages. Local farmers live primarily on cropland cultivation, animal husbandry, and forest resource extraction. They also engage in other economic activities such as local off-farm employment, local businesses, and out-migration.

6.2 Concurrent PES programs in the Tianma National Nature Reserve

The GTGP is the largest PES program in the world. It was implemented in the aftermath of back-to-back natural disasters in the late 1990s in China. The program

DOI: 10.4324/9781003290292-6

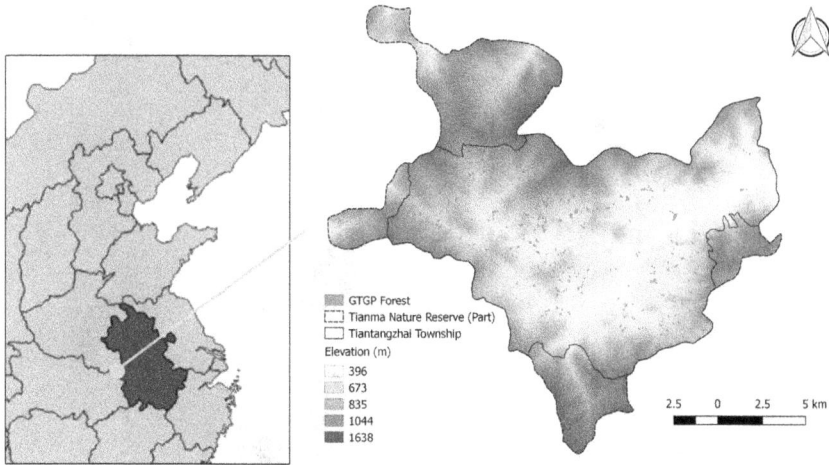

Figure 6.1 Location of Tianma National Nature Reserve (TNNR). The core zone is situated within Tiantangzhai Township, Anhui Province, China.

was initially tested in Shaanxi, Shanxi, and Sichuan provinces before becoming a national policy in 2001 (Zhang & Song, 2006). By enrolling in the GTGP, farmers plant trees in croplands on steep slopes for soil and water conservation, and the Chinese central government provided the farmers with as much grain as they would harvest from the cropland as compensation. Thus, the program is named the GTGP, although the grain compensation was replaced with cash in the subsequent years due to high transaction costs. Due to the rough terrain, some croplands in Tiantangzhai are located on steep slopes, causing severe soil erosion. Therefore, the GTGP was applicable in this region and was implemented in 2002, resulting in the enrollment of 1,680 mu (112 ha) croplands from 753 households. Most enrolled lands in the GTGP in Tiantangzhai were established as ecological forests, mainly sweetgum (*Liquidambar styraciflua*). The duration of payment was eight years for ecological trees and five years for economic trees (e.g., pecan trees) according to the GTGP policy. The compensation rate was set the same as in the Yangtze River Basin (Song et al., 2014), i.e., 210 yuan/mu/year during the first eight-year contract (ecological trees) and then 125 yuan/mu/year for the renewed eight-year contract. Under the supervision of the local government, qualified croplands were first identified for enrollment, and then the households that farm these croplands were requested to enroll in the program. Theoretically, the farmer can decline the request, but it rarely happens in reality.

The FEBC program in TNNR was initially embedded in a forest management policy (Dai et al., 2009), called Ecological Welfare Forest Program, aiming to protect natural forests from commercial logging (Zhang et al., 2019). The nature of the program is similar to the national forest conservation policy, the Natural Forest Conservation Program. In 2000, the Chinese government adopted the PES

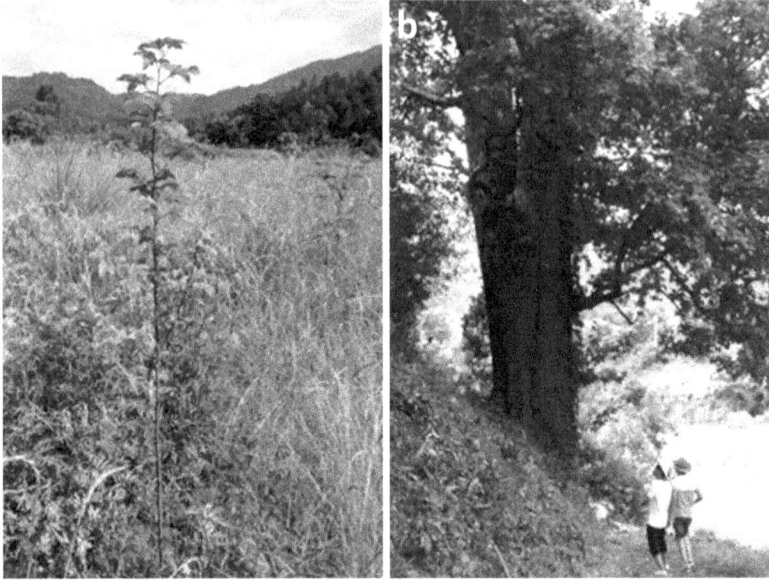

Figure 6.2 Photos of tree saplings and forests protected under the FEBC. (a) Tree saplings recently planted on cropland enrolled in the GTGP and (b) protected natural forests under the FEBC program in the Tianma National Nature Reserve.

principle and compensated households who own FEBC forests. The compensation rate in Tiantangzhai was initially set at 5.00 yuan/mu/year, and the compensation rate increased to 8.75 yuan/mu/year in 2009. Since most households have some natural forests in the study area, nearly all households automatically participated in the FEBC and received government payments (Figure 6.2). The area of FEBC land varies widely among households due to the natural variation in forest areas from village to village. Those living in the mountains often have extensive natural forests and are compensated with sizable cash payments each year. According to our survey, the total payment amount for a household can be as high as several thousand Chinese yuan per year, which may account for the majority share of total agricultural income for the household (Zhang et al., 2018c).

6.3 Data collection

We conducted two waves of household surveys collecting data to evaluate the socioeconomic and ecological impacts of the GTGP and the FEBC in Tiantangzhai Township in 2013 and 2014. Since nearly every household is enrolled in the FEBC program, we designed our sampling schemes focusing on the GTGP. In the 2013 survey, we adopted a simple random stratified sampling scheme. We had two household strata, the households enrolled in GTGP and those not enrolled. We aimed to collect a roughly equal number of households from each stratum. We successfully interviewed 250 households comprising 139 GTGP participating households and 111 non-participating households in 2013. We obtained information on demography, economic activities, land use, and compensation from each of the PES programs from each interviewed household. We particularly asked whether people stole any trees from FEBC land (i.e., FEBC tree theft) and whether a local forestry station monitored FEBC forests. For each of the interviewed households, we also visited every cropland parcel the household farmed at the time and recorded its geolocations with a hand-held GPS unit. The geographic coordinates allow us to derive the topographic features (e.g., slope and aspect) with a digital elevation model (DEM) and connect the biophysical data at the cropland parcel level with the household socioeconomic data. We visited 1,196 cropland parcels in the 2013 household survey.

In the 2014 household survey, we designed a more comprehensive questionnaire because we received funding from the US National Science Foundation (grant number DEB-1313756), covering more topics than that used in the pilot study in 2013. We adopted a more sophisticated sampling method, a two-stage disproportionate random sampling scheme (Bilsborrow, 2016), to select households for the survey. Since many more households were not enrolled in the GTGP, and we aimed to select a sample with a roughly equal number of enrolled and those not enrolled in the GTGP, we need to oversample in the GTGP participating households and under-sample in the GTGP non-participating households. We first obtained a list of the household population of Tiantangzhai, including information on household head name, resident group, village, and whether they are participating in the GTGP. In the first stage, we selected resident groups based on the proportion of households enrolled in the GTGP, which was stratified into five strata, with their participation proportion being 1.0–0.80 (Stratum-I), 0.79–0.50 (Stratum-II), 0.49–0.30 (Stratum-III), 0.29–0.00 (Stratum-IV), and 0.00 (Stratum-V). We randomly selected ten, nine, seven, ten, and four resident groups from each of the strata, resulting in an estimated mean proportion of GTGP participation of 47%. In the second stage, we randomly selected roughly 20 households from each resident group from the two strata of GTGP participants and non-participants. If a resident group consists of fewer than 20 households, all households would be selected. We purposely oversampled GTGP participating households for resident groups with lower proportions of participation in the GTGP and oversampled non-participants for resident groups with a higher proportion of households participating in the GTGP. This sampling would lead to a final sample with a balanced number of households for GTGP participants and non-participants from each resident group. We successfully collected data for 481 households, with 271 (56%) participating in the GTGP and the remaining households not enrolled in the GTGP. A more detailed description of the sampling process can be found in previous studies (Song et al., 2018; Zhang et al., 2020).

In both surveys, we obtained the geolocation information of each interviewed household with a hand-held Global Positioning System unit. Ancillary data, including a high spatial-resolution remotely sensed image (WorldView-2 on 7/13/2013), a digital elevation model, and topographic maps, were used to delineate each of the GTGP forest stands in Tiantangzhai. Given these data, we performed the following analyses concerning the spillover effects among the two concurrent PES programs, the GTGP and the FEBC.

6.4 Data analysis and modeling

With careful data collection, this section explores five fundamental dimensions of local livelihoods: cropland abandonment, household labor allocation (for labor out-migration), household energy transition (from fuelwood to nontraditional, alternative sources), tree theft, and the direct impact from FEBC to GTGP. In particular, we focus on how the two green initiatives, GTGP and FEBC, may interact with each other and account for the above four dimensions.

6.4.1 Cropland abandonment

Cropland abandonment has been a significant phenomenon in mountainous areas (Zhang et al., 2014), like Tiantangzhai Township in China. The abandonment of cropland potentially contributes to the additionality of the PES programs for ecological restoration. It is a reverse process of conversion from natural surface to human-dominated land. The GTGP and the FEBC may change rural households' land-use decisions on cropland abandonment through two mechanisms. First, the cash compensation provides households with financial resources to reallocate farm labor to other activities than cropland cultivation, causing some cropland parcels in marginal areas to be abandoned due to lack of labor. Second, the converted forests under the GTGP and the recovery of natural forests under FEBC can influence the decision-making on cropland use located near the forests due to multiple feedback effects such as crop raiding by wildlife (Chen et al., 2019) and shading effects (Bista et al., 2021). However, the compensation from the GTGP and the FEBC provides financial resources to purchase agricultural tools and supplies, which may lead to agriculture intensification. It is essential to understand the respective effect of the two PES programs and their ensemble effects on cropland usage.

We first used the data of the 250 households from the 2013 household survey because the survey contains the geolocation data for 1,196 cropland parcels managed by these households. Although households are central decision-makers for cropland use, the abandonment of a cropland parcel also depends on its biophysical conditions. Therefore, the key factors influencing cropland abandonment include the biophysical features of the cropland parcel and the socioeconomic and demographic characteristics of the household to which the cropland parcels belong (Table 6.1). By setting the dependent variable as a binary variable indicating whether a cropland parcel has been abandoned (0 = cultivated, 1 = abandoned), we utilized a random-coefficient modeling approach to examine the effects of PES programs on cropland use decisions, with the equation as follows:

Table 6.1 Explanatory variables for modeling cropland abandonment at the TNNR, China

Variable	Description	Mean (std. dev.)
Parcel level		
Parcel area	Area of land parcel (ha)	0.0861 (0.0841)
Parcel type	If the parcel is dryland (0 = no, 1 = yes)	0.4534 (0.4980)
Walk distance	Reported walking time from parcel to house (minutes)	10.6199 (10.6994)
Parcel elevation	Elevation at parcel location (100 m)	6.466 (0.9646)
Parcel TWI	Topographic wetness index value at parcel location	9.8707 (3.9722)
Parcel aspect	Aspect facing the direction of the parcel (0 = south, 180 = north)	75.6614 (52.7316)
GTGP distance	Geographic distance to nearest GTGP forest stand (100 m)	0.7849 (0.7269)
FEBC distance	Geographic distance to nearest FEBC natural forest edge (100 m)	3.4408 (3.1647)
Household level		
Head age	Age of household head	52.2238 (9.2554)
Head gender	Gender of household head (0 = male, 1 = female)	0.0341 (0.1816)
Head education	Education completed by the household head (years)	6.9642 (2.6433)
House elevation	Elevation at house location (100 m)	6.4308 (0.9427)
Farm labor	Number of a current household member aged 18–60, living at home, and being able to provide farm labor	1.8319 (1.0751)
Cropland	Total area of cropland managed (ha)	0.4146 (0.1854)
Crop raiding	Incidence of crop raiding by wildlife (0 = no, 1 = yes)	0.4634 (0.4989)
Livestock	Livestock ownership (0 = no, 1 = yes)	0.7696 (0.4213)
Off-farm	Share of local off-farm income in total household gross income	0.2823 (0.4016)
Fuelwood use	Total amount of fuelwood used per year (1,000 kg)	9.2003 (5.7606)
FEBC	Total amount of payment received from FEBC per year (1,000 yuan)	0.3349 (0.3581)
GTGP	Total amount of payment received from GTGP per year (1,000 yuan)	0.1583 (0.2327)

$$y^* = \alpha + \beta X + \gamma Z + \mu + \varepsilon \qquad (6.1)$$

where y^* is the transformed logistic indicator denoting whether a parcel is abandoned or not; X is the vector of parcel-level variables and Z is the vector of household-level variables; α is the intercept; β and γ are fixed effects corresponding to parcel-level and household-level variables, respectively; μ is the random coefficient that can capture the household-level variance; and ε is the random error term at the parcel level.

6.4.2 Household labor allocation for labor out-migration

Rural-to-urban migration is a hallmark of the socioeconomic transformation in China following the adoption of the open and reform policy in the late 1970s (Peng, 2011). Currently, there are about 200 million migrants from rural areas working in cities in China, pulling hundreds of millions of people out of poverty in China (Liang, 2016). Tiantangzhai experienced a growing trend of out-migration when the FEBC and the GTGP were implemented. According to theories in micro- and macroeconomics, migration can be viewed not only as an individual-level decision-making outcome but also regarded as a livelihood strategy of a household (Barbieri et al., 2009; Bilsborrow, 2016; Bilsborrow et al., 2004; Massey, 1990).

The intervention of PES programs can influence the livelihood strategy of households who enrolled their lands in the programs. Farmers used to be intimately connected to their land, which produces livelihood necessities, and active farming secures land tenure simultaneously (Ma et al., 2015). On the one hand, the two PES programs, especially the GTGP, directly change land-use types from cropland to forest, relaxing the labor liquidity constraints, and hence indirectly support the livelihood diversification (Lin & Yao, 2014). Furthermore, the cash compensation can serve as a safety net for risk diversification by investing the labor force into multiple alternative off-farm activities, particularly migration, which can bring in lucrative economic returns from remittance (Zhang et al., 2019). On the other hand, households may also invest financial capital received (i.e., cash compensation) to intensify agriculture by using more fertilizer and/or allocating more labor time to land cultivation after enrolling marginal croplands in the GTGP; on the contrary, the effects of the payment schemes between FEBC and GTGP may be different in influencing household livelihoods. Here, we analyze how the two concurrent PES programs affect individual migration.

We draw on socioeconomic and demographic data from the 2014 household survey to examine the effects of the current PES programs on labor migration. We define a migrant as an individual who lives away from the household outside the county for at least six consecutive months and is 15–59 years old at the survey time. Individuals outside this age range are more likely to be dependent and thus not considered migrants in this study. We recorded the migration history of each individual from 2000 to 2013 in the interviewed households, and the migrants can be easily identified from the interview data recorded. For time-varying variables such as age and education, we reconstructed panel data to reflect the status of migrants before the migration. We also randomly selected a non-migrant from each interviewed household and obtained their attributes (e.g., age, education, and occupation) when the migrant left the household. For households with no migrant, we recorded a randomly selected house member and his or her attributes five years before the survey time, which is roughly the midpoint of our study period, making the non-migrant group comparable to the migrant group. The dependent variable is the migration status for all individuals aged 15–59 (1 = yes, 0 = no).

We used logistic regression to analyze factors that influenced the migration decision. To control for the contextual factors, we developed a multilevel regression model (Equation 6.2), including individual attributes (*I*), household characteristics (*H*), and resident group factors (*G*), to understand the factors influencing individual migration (*M*) (Table 6.2). We mainly included payments received from the FEBC and the GTGP at the household level to examine their effects on migration.

$$M = f(I, H, G) \tag{6.2}$$

Table 6.2 Statistics of explanatory variables for modeling labor migration at TNNR, China

Variable	Description	Mean	Std. dev.
Individual level			
Gender	Individual gender (0 = male, 1 = female)	0.4969	0.5002
Age	Individual age	36.6676	13.2997
Education	If individual completed elementary school (0 = no, 1 = yes)	0.7573	0.4289
Marriage	Marital status of individual (0 = never married, 1 = married)	0.7274	0.4455
Single female	If individual is a single female (0 = no, 1 = yes)	0.1108	0.3140
Household level			
Head gender	Gender of household head (0 = male, 1 = female)	0.0457	0.2090
Head age	Age of household head	49.6992	9.1721
Head education	If household head completed elementary school (0 = no, 1 = yes)	0.7537	0.4310
Head marriage	Marital status of household head (0 = never married, 1 = married)	0.9120	0.2833
Elevation	Elevation at house location (m)	673.5856	104.0458
Walk	Travel time to nearest paved road by walk (minute)	11.3773	14.4167
Household size	Number of current household members	3.7168	1.2308
Migration experience	If any household member or previous member has migration experience (0 = no, 1 = yes)	0.3219	0.4674
Cropland	Total amount of cropland under cultivation (mu)	4.8173	3.3076
GTGP	Payment amount received from GTGP per year (1,000 yuan)	0.1486	0.1993
FEBC	Payment amount received from FEBC per year (1,000 yuan)	0.4979	0.6243
Group level			
Group size	Number of households within the resident group	26.0185	8.6601
School	Distance to nearest elementary school (minute)	20.0545	25.0469
Hospital	Distance to nearest hospital or clinic (minute)	18.7599	15.8205

6.4.3 Energy transition: fuelwood vs. alternative sources

Many rural regions in developing countries like China are still at the early stage of energy transition (Tang & Liao, 2014). Fuelwood collected from natural forests remains the primary energy source for households living in forest areas (Zhang et al., 2009). At TNNR, fuelwood is used for cooking daily meals and livestock feed and heating during the winter (Song et al., 2018). According to the energy ladder theory (Leach, 1992), a household tends to go through a transition from using primitive (e.g., fuelwood) to transitioning (e.g., kerosene or coal) and to modern fuels (e.g., electricity or Liquid Petroleum Gas) as the household income increases. The fuel stacking theory posits that a household adopts new fuels as income increases without altogether forgoing the old fuels, i.e., households do not switch fuels but "expand" their fuel portfolio as income increases (Masera & Navia, 1997). The change in energy sources from fuelwood to modern energy sources aligns with the goal of forest conservation policies because the transition relaxes the pressure on forest resources. The compensation from the two PES programs, particularly the FEBC program, reduces rural households' dependence on their land and encourages them to seek alternative livelihoods, potentially influencing rural household energy uses conducive to forest sustainability.

We used the 2014 household survey data to examine how PES affected rural households' fuelwood use and fuel choices. We designed two sets of questions relating to energy use in the questionnaire. The first is fuel choices, including (1) using fuelwood or other biomass (e.g., crop stalk) as the only energy source; (2) using fuelwood as the primary energy source, supplemented with modern fuels; (3) roughly half and half of fuelwood and gas/electricity as sources for energy; (4) using gas/electricity as the primary energy source, supplemented with fuelwood; and (5) using gas/electricity as the only energy source. The second set aimed to quantify the amount of fuelwood used for cooking, heating, and feeding per year. To lower the burden of estimation by the respondents, we asked them to estimate the quantity per day and then computed the amount of usage per year by the interviewers after the interview. The sum of the usage for the three activities was the total fuelwood use. By controlling for various socioeconomic factors, we fitted a multinomial logistic regression model and a weighted multiple regression model to understand the factors affecting fuel choices (categorical) and per capita fuelwood use (continuous) in rural households. The models for fuel choice and fuelwood amount are respectively specified as follows:

$$ln\left[\frac{\Pr(Y = j \mid X)}{\Pr\left(Y = i|X\right)}\right] = \beta X + \varepsilon \tag{6.3}$$

$$U = \gamma X + \varepsilon \tag{6.4}$$

where Y represents the household fuel choices (j, $j \neq i$; i is the reference fuel choice, which is set as fuelwood/coal as the only energy source); X is the vector of explanatory variables, while β captures the fixed effects of the explanatory variables; U represents the quantity of per capita fuelwood use in a household, while γ captures the fixed effects corresponding to X, and ε represents random errors. Note that in the first equation, we merged fuel choices (4) and (5) into one choice (using gas/electricity as the primary or only energy source) because very few households used gas/electricity as the only source of energy.

6.4.4 Tree theft

Despite the overall increase in forest cover since implementing the PES programs at TNNR (Zhang et al., 2018b), tree theft has also been reported by rural households from the 2013 household survey. This finding is another spillover effect of restricting forest resources under the two PES programs, particularly the FEBC program. Rural households enrolled in FEBC cannot harvest timber from the natural forests they manage. However, the compensation may not effectively relax their dependence on forest resources, such as fuelwood or other wood product need. Thus, households may have to extract resources from forests in surrounding areas managed by others, resulting in a negative spillover of deforestation or forest degradation, jeopardizing the effectiveness of forest conservation under the PES programs. Thus, we aimed to understand if the GTGP and the FEBC played a role in this spillover effect of tree theft.

Our 2013 household survey data recorded the tree theft information, thus enabling us to investigate whether tree theft on FEBC land is related to GTGP after controlling for other factors. We hypothesized that tree theft on FEBC land might be affected by the FEBC land area, GTGP participation, the closeness of houses to the GTGP forests, and neighbors' GTGP forest area. The variable of neighbors' GTGP forest area was defined as the GTGP area enrolled by neighboring households within the same resident group, i.e., the total GTGP area in the resident group subtracting the GTGP forest area of the household of interest. Since neighbors' GTGP forest area is a group-level factor, we developed a mixed-effects model that captures random effects at household and resident group levels to test our hypotheses.

6.4.5 Direct interactions between GTGP and FEBC

In addition to the complex spillover effects described in the previous sections, we examined the direct interactions between the GTGP and the FEBC to understand synergistic or offsetting effects. The enrollment in the FEBC was pre-registered before 2002 as a continuing effort of the forest conservation policy (Dai et al., 2009), whereas the GTGP was newly initiated and implemented in 2002. Hence, participation in the FEBC program can be considered a pre-existing condition that may influence the enrollment in GTGP by rural households in TNNR.

We used data from household surveys in 2013 and 2014 to investigate the interaction between the GTGP and the FEBC. Since the spillover effect between the two concurrent programs is of interest, we excluded households that did not participate in both PES programs from the sample. This way, we obtained a subsample of 408 households, including 139 and 269 households, from the 2013 and 2014 surveys, respectively. We developed a multivariate linear regression model to examine the relationship between the cropland area enrolled in the GTGP, and the forest area enrolled in the FEBC, controlling for other factors such as household demographic and socioeconomic conditions. Households with large areas of FEBC forests tend to live in high elevations and thus may naturally manage croplands that are more likely to be targeted for GTGP enrollment. Therefore, we also included household elevation in the model to control its confounding effects.

6.5 Findings

With the above data analysis and modeling efforts presented, we show the results in the same order: cropland abandonment, household labor allocation for outmigration, energy transition: fuelwood vs. alternative sources, tree theft, and direct interactions between the GTGP and the FEBC.

6.5.1 Cropland abandonment

We found that the geographic locations of the GTGP and the FEBC forests relative to the cropland parcels exhibit statistically significant effects on cropland abandonment, but only the FEBC payment influences household decisions on cropland abandonment (Figure 6.3a). Every 100 m increase in distance of a cropland parcel to the nearest GTGP or FEBC forests decreases the odds ratio of abandoned versus not by 8% and 46%, respectively. This result means that cropland parcels in proximity to FEBC and GTGP forests are more likely to be abandoned, and the effect of the FEBC forests is much stronger than that of the GTGP on cropland abandonment. The FEBC forests are natural forests where more wildlife likely resides. Wildlife can cause significant damage to the crop, contributing to farmers' decision to abandon the cropland (Bista & Song, 2021).

After controlling for a series of other socioeconomic variables (Figure 6.3b), we found that the GTGP payment, which is relatively small in amount, did not significantly affect household cropland use decision on abandoning cropland parcels. However, every additional 1,000 yuan of a household's FEBC payment can increase the cropland abandonment odds ratio by over two times. The FEBC pays approximately three times more cash compensation to the participating households than the GTGP pays (Song et al., 2018), making the households receiving more compensation from the FEBC affordable to abandon the marginal croplands. By inducing more cropland abandonment, the two PES programs may synergize the additionality of environmental restoration. The abandoned cropland parcels will go through secondary succession and eventually become forested lands, constituting positive spillover effects in line with the aims of the two PES programs.

(a) Cropland parcel level

(b) Household level

Odds Ratio

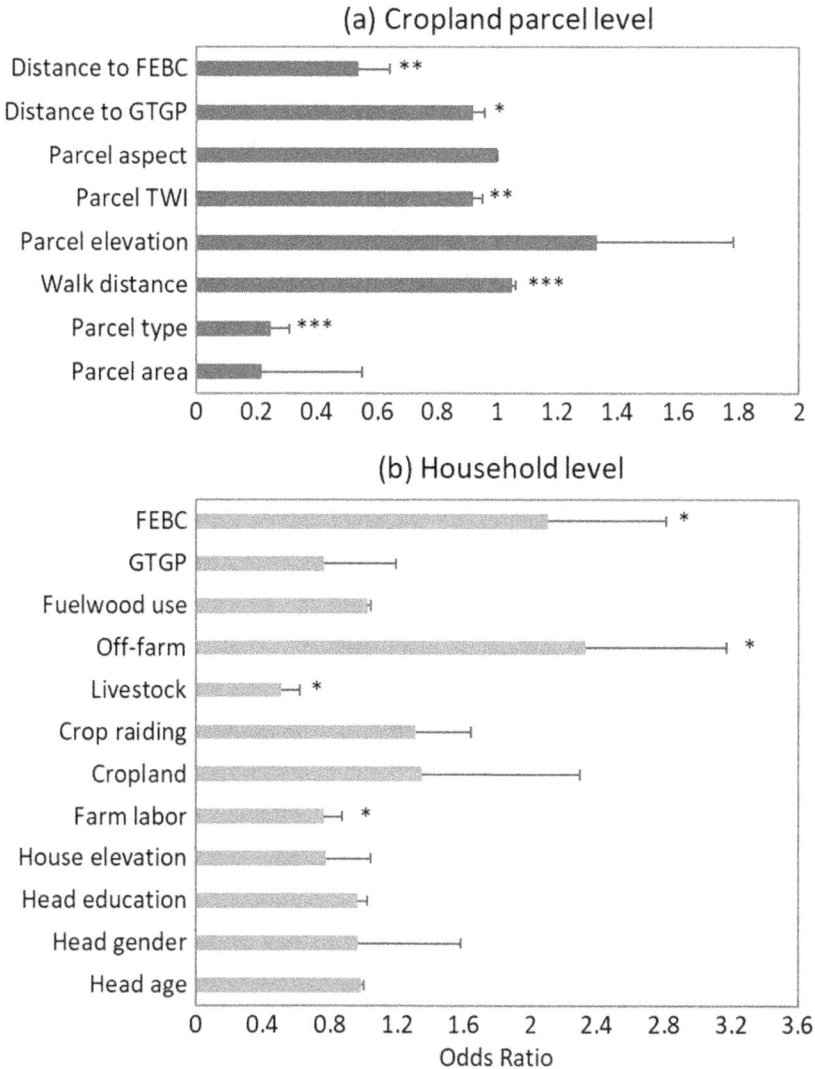

Figure 6.3 Effects of the GTGP and the FEBC on cropland abandonment by farm households. The effects are based on random-coefficient models at TNNR, China. (a) Modeling result at individual cropland parcel level; (b) modeling result at the household level. The model uses data collected from the 2013 survey with 1,196 cropland parcels managed by 250 households. $*p<0.10$; $**p<0.05$; $***p<0.01$.

6.5.2 Household labor allocation for out-migration

Results from the multilevel analysis suggest negative spillover effects between the two PES programs. The logistic regression analysis found that the GTGP and the FEBC had opposite effects on labor out-migration after controlling for other factors (Table 6.3). Specifically, every additional 1,000 yuan of GTGP payment increased the odds of sending out migrants by 4.4 times, consistent with classic PES literature regarding GTGP's positive impact on out-migration (Uchida et al., 2009). On the other hand, every additional 1,000 yuan of FEBC payment decreased the odds of out-migration by 34%.

The GTGP seeks to enroll marginal cropland for reforestation, thus freeing farm labor from land cultivation. Without involving cropland retirement, the FEBC provides sizable cash compensation to participating households, particularly those with large areas of FEBC forests (Zhang et al., 2018c). The significant compensation from the FEBC with nearly minimal opportunity cost can

Table 6.3 Results of the multilevel logistic regression model of migration decisions at Tianma, China

Variable	Coefficient	Odds ratio	Robust standard error
Individual level			
Individual gender (1 = female, 0 = male)	−1.4318***	0.2389	0.0631
Individual age	−0.1528***	0.8583	0.0150
Individual education	1.1852**	3.2714	1.9021
Individual marital status (1 = married, 0 = otherwise)	1.3118**	3.7128	2.2346
If individual is a single female (1 = yes, 0 = no)	1.1314**	3.1001	1.7288
Household level			
Gender of head	−1.4776	0.2282	0.2315
Age of head	0.1264***	1.1347	0.0297
Education of head	0.4410	1.5543	0.5446
Marital status of head (1 = married, 0 = otherwise)	−1.0293	0.3573	0.3417
House elevation (m)	0.0037	1.0037	0.0048
Walking distance to nearest paved road (minute)	−0.0337	0.9669	0.0232
Household size	−0.0545	0.9470	0.0880
If household has previous migration experience (0/1)	1.2267***	3.4098	1.1259
Cultivated land area (mu)	−0.2016***	0.8174	0.0372
GTGP payment (1,000 yuan)	1.6870***	5.4031	2.9277
FEBC payment (1,000 yuan)	−0.4116*	0.6626	0.1535
Resident group level			
Resident group size	0.0015	1.0015	0.0180
Distance to nearest elementary school (minute)	0.0065	1.0065	0.0049
Distance to nearest hospital or clinic (minute)	0.0151**	1.0152	0.0067
Intercept	−3.4330	0.0323	0.1474
Intercept variance	0.4943		0.3729

Notes: *$p<0.10$; **$p<0.05$; ***$p<0.01$. The model uses data collected from a 2014 survey with a sample of 1,137 individuals from 412 households in 40 resident groups. The results are from Zhang et al. (2018a).

substantially improve the livelihoods of the participating households, reducing the need for income from out-migration. Understandably, the FEBC compensation may have lessened the pressure of cash shortage for the enrolling household, allowing household members to stay together with advantages for caring for the elderly, children's education, and quality family life. The GTGP has strongly stimulated residents to become out-migrants, releasing local population pressure on natural resources, while the FEBC has the opposite effect on out-migration.

6.5.3 Energy transition: fuelwood vs. alternative sources

The distributions of fuel choices and fuelwood use quantity indicate that rural households at TNNR are still in the early stage along the energy ladder, predominantly using fuelwood for energy (Figure 6.4). Among all the interviewed households, over 70% reported that they used fuelwood as the primary energy source, and about 10% of the households interviewed used fuelwood as the only energy source. The percentages of households selecting the other sources as primary sources of energy are much lower: only 8% and 6% fall in Category 3 (roughly half fuelwood and half modern fuel) and Category 4 (gas/electricity as the primary source of energy), respectively, whereas a trivial number (2%) used modern fuels as the only source of energy. On average, the total amount of fuelwood consumption is as high as 10,147 kg per year; cooking for daily meals and heating during winters are the two major activities for fuelwood consumption.

Based on the modeling results (Table 6.4), we found that only the forest area enrolled in the FEBC program had a statistically significant effect on fuel choices and fuelwood use, but the GTGP did not significantly affect either fuel choice or fuelwood amount used. Households with larger areas of FEBC forests are more likely to retain fuelwood as the only source of energy compared to other options. We also found that every mu (~667 m^2) of FEBC forests would increase the quantity of per capita fuelwood use by 12.1 kg, making the household more dependent on forest resources. The two PES programs did not seem to substantially shift the daily use of fuel of the participating households from fuelwood to cleaner modern fuels because fuelwood is accessible to them with plenty of supply in the study area. New policies specifically designed to change farmers' behavior from using fuelwood to cleaner modern fuels are needed if policymakers aim to reduce fuelwood use and preserve forest resources.

6.5.4 Tree theft

Among the 250 surveyed households in 2013, 32% reported that they experienced tree theft in the natural forests they managed (Table 6.5). According to the model on tree theft on FEBC land, every 100 m closer to a household residence to the nearest GTGP land increased the odds of FEBC tree theft by 15.5% (i.e., $1 - \exp(-0.1685)$) after controlling for other socioeconomic factors (Table 6.6), suggesting that trees on FEBC land are more likely to be illegally logged by neighboring residents if the household is in closer proximity to GTGP land. Such

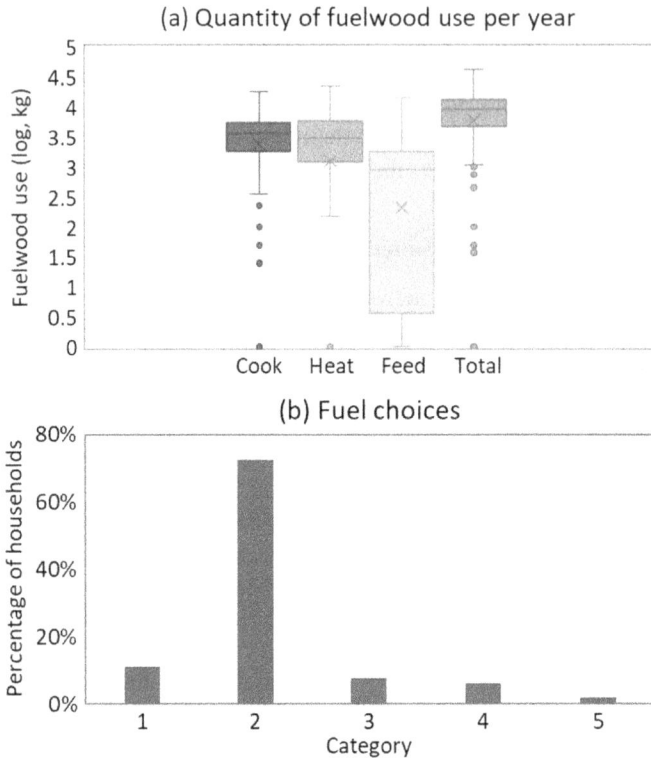

Figure 6.4 The statistical description of fuelwood use and fuel choices. (a) Distributions of fuelwood use amount for cooking, heating, feeding, and all activities; (b) distribution of households with different fuel choices (1: fuelwood or coal as the only energy source; 2: fuelwood or coal as the primary energy source, supplemented with modern fuel; 3: approximately half fuelwood/coal and half gas/electricity for energy; 4: gas/electricity as the primary energy source, supplemented by fuelwood; 5: gas/electricity as the only energy source). The unit is the log-transformed kilogram. Data are derived from the 2014 household survey.

a tree theft phenomenon is an example of a *Policy–Behavior* spillover effect, i.e., the payment from GTGP (Policy 1) may lead to an unintended behavior of tree theft on FEBC land (Behavior 2).

This *Policy–Behavior* spillover effect may arise from a *Behavior–Behavior* spillover effect: migration of the whole family or farm laborers to cities for higher-paying employments (Behavior 1) may lead to a reduction or cessation of monitoring the FEBC forests belonging to the household, giving the perpetrators the chances to steal trees from these forests (Behavior 2). These findings reflect a hidden, negative spillover effect from the GTGP to the FEBC. Given that the GTGP actively promotes out-migration and out-migration increases the probability of tree theft, GTGP may indirectly lead to tree theft, thus degrading

Table 6.4 Results of modeling fuel choices by households in TNNR, China

Variable	Fuel choice			Fuelwood quantity used per capita
	Score-4	Score-3	Score-2	Coef. (std. err.)
Age of oldest household member	0.977**	1.033***	0.988*	41.2 (16.0)**
Education of household head	1.027	1.022	0.977	−138.3 (58.2)**
Wellness index	1.946***	1.203***	1.152***	−32.9 (41.3)
Household income (natural log)	5.734***	3.698***	2.421***	−386.2 (192)**
Walking time from home to main road (minute)	0.823***	0.957***	0.978***	9.5 (11.6)
Household size	0.938	0.698***	0.61***	−1,146.5 (133.0)***
GTGP area (mu)	0.92	1.022	1.108	120.4 (165.4)
FEBC area (mu)	0.990***	0.974***	0.992***	12.1 (3.6)***
Paddyland under cultivation (mu)	0.757***	0.907**	1.101***	−54.1 (57.2)
Dryland under cultivation (mu)	0.37***	0.558***	0.839***	260.5 (126.6)**

Notes: Fuel score 1 is for households using fuelwood or other solid fuel as the only source of energy. Fuel score 2 is for households using fuelwood or other solid fuel as the primary source of energy with modern fuel as supplementary. Fuel score 3 is for households using half fuelwood or other solid fuel and half modern fuel. Fuel score 4 is for households using modern fuel as the primary or sole source for energy. $*p < 0.1$; $**p < 0.05$; $***p < 0.01$.

Table 6.5 Statistics of summary explanatory variables for modeling tree theft at TNNR, China[a]

Variable	Mean	Std. dev.	Min	Max
Reported tree theft (0 = no,1 = yes)	0.3160	0.4658	0	1
Neighbors' GTGP area (mu)	16.5565	15.9859	0	60.1000
Geographic distance from residence to nearest GTGP land (100 m)	3.4358	2.9849	0.0116	18.6297
Participation in GTGP (0 = no, 1 = yes)	0.5560	0.4979	0	1
FEBC area (mu)	37.7295	46.2826	1	350
FEBC monitoring (0 = no, 1 = yes)	2.0160	0.5806	1	3
Household elevation (100 m)	6.4488	0.9971	4.0500	8.7500
Household head's age	52.4440	9.6159	31	78
Household head's gender (0 = male, 1 = female)	0.0480	0.2142	0	1
Household head's education	6.9560	2.7099	0	14
Household head an out-migrant (0 = no, 1 = yes)	0.2360	0.4255	0	1
Household size	4.5800	1.3752	1	9
Cropland area owned (mu)	5.7128	2.7062	0	16.1000
Livestock ownership (0 = no, 1 = yes)	0.8520	0.3558	0	1
Fuelwood use (1,000 kg)	8.8202	5.9169	0	36.2500
Off-farm income (1,000 yuan)	58.0244	78.8633	0	730

Note: [a] The model is for modeling FEBC tree theft from the 2013 household survey.

Table 6.6 Results of mixed-effects logistic regression of FEBC tree theft at TNNR, China

Variable	Coefficient	Standard error
Neighbors' GTGP area (mu)	0.0072	0.0136
Geographic distance of household to nearest GTGP land (100 m)	−0.1685**	0.0759
GTGP participation (0 = no, 1 = yes)	−0.5651	0.4180
FEBC area (mu)	−0.0013	0.0040
FEBC monitoring (0 = no, 1 = yes)	0.1924	0.2699
Household elevation (100 m)	−0.2457	0.2170
Household head's age	0.0189	0.0179
Household head's gender (1 = female, 0 = male)	1.1148	0.6995
Household head's education	−0.0432	0.0632
Household head migration status (0 = no, 1 = yes)	0.8855**	0.3796
Household size	−0.0056	0.1190
Cropland owned (mu)	−0.0049	0.0638
Livestock ownership (0 = no, 1 = yes)	−0.1191	0.4537
Fuelwood use (1,000 kg)	−0.0042	0.0306
Off-farm income (1,000 yuan)	0.0019	0.0020
Constant	0.2686	1.9845
Constant variance	0.2577	0.4198

Notes: The model uses data collected from the 2013 survey with a sample size of 250. The dependent variable is the occurrence of tree theft (0 = no, 1 = yes). Neighbors' GTGP area is calculated as the total GTGP area of the resident groups minus the GTGP area of the household of interest in this resident group.

the FEBC forests. Meanwhile, households close to GTGP lands tend to live in areas with a harsh geographic environment and thus depend more on forest resources for livelihoods. In contrast, households that are further away from GTGP lands have better opportunities to engage in alternative livelihoods and diversify income sources, being less dependent on timber and fuelwood from FEBC forests. Moreover, households with more extensive FEBC forests rely more on fuelwood use for energy, further compromising forest conservation effectiveness.

6.5.5 Direct interactions between the GTGP and the FEBC

Based on the combined household sample from 2013 and 2014, we found that households enrolled much more land in the FEBC (50.12 mu ± 62.74) than in the GTGP (2.16 mu ± 1.73) on average, and the distribution by the enrolled area in the GTGP among the enrolled households is more even than that of the FEBC (Figure 6.5). The majority of participating households (81%) enrolled 0.05–3 mu of cropland in the GTGP, whereas nearly three-quarters enrolled less than 50 mu of forestland in the FEBC program. We found both positive and negative hidden

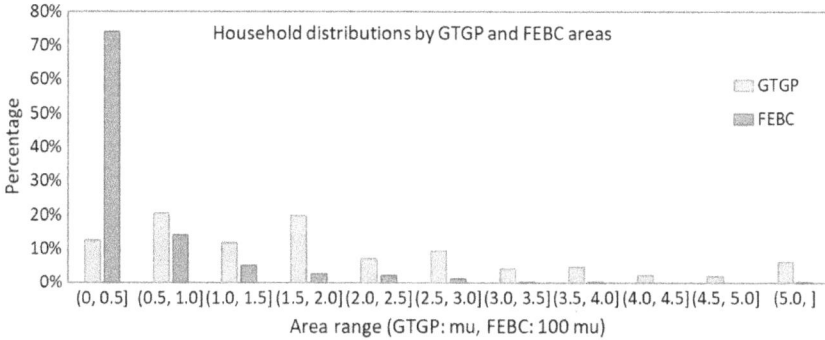

Figure 6.5 Distributions of households by the areas of land enrolled in GTGP and FEBC. Note that the unit of the FEBC area is 100 mu, so the mean FEBC area is generally two magnitudes larger than the mean GTGP area. 1 mu = 1/15 ha. Data are derived from the 2013 and 2014 surveys at the TNNR, China.

spillover effects between the two programs according to the regression model. Cropland area enrolled in the GTGP (dependent variable) is significantly positively associated with FEBC forest area (coefficient = 0.4694, p = 0.002; Table 6.7) after controlling for other factors, indicating that every 100 mu of FEBC forestland leads to an additional 0.47 mu of cropland enrolled in the GTGP.

The above positive *Policy–Behavior* spillover effect may come from local farmers' adaptive livelihood strategy. After receiving FEBC payments, the recipient households are required to refrain from timber harvesting and some limited responsibilities in fire prevention and anti-theft patrol. Households with larger FEBC areas (i.e., receiving large FEBC payments) tend to have more cropland parcels located in marginal areas on steep slopes. When compensation comes from FEBC, local households may afford to reduce their farming activities in these marginal areas, enrolling them in GTGP. Despite a relatively low compensation rate of FEBC on the unit area basis (131.25 yuan/ha in 2014 in Tiantangzhai) compared with that of GTGP, the average total compensation received from FEBC was approximately three times that from GTGP due to the large areas of natural forests belonging to households. Some local farmers may afford to buy more food or fodder from the local market, increasing their confidence in food security, making a comfortable living with the income from FEBC alone and/or some local off-farm employment. This situation may make local farmers more willing to enroll marginal (e.g., distant land on steep slopes) cropland parcels in the GTGP. Local farmers may further increase the household income with the freed labor from farming the GTGP land switched to local off-farm employment or migration to cities for higher wages. Unlike the GTGP, the FEBC does not reduce the cropland area from farming, meaning it does not directly free farm

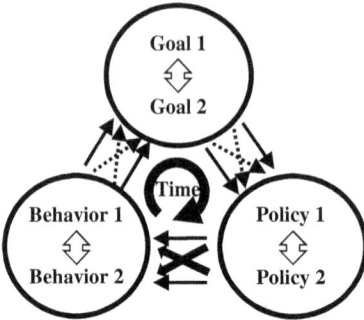

Figure 6.6 Cross-program spillover effects at the TNNR, China. This diagram is modified
from Figure 1.3, where the solid one-way arrows stand for internal influences
from one element to another within the same initiative, while the dashed one-
way arrows and double two-way arrows for potential spillover effects; the
circular one-way arrow represents Time–Time spillover effects. The shaded,
bold arrow represents the spillover effect with evidence from this section.

Table 6.7 Results of regression of GTGP area against FEBC area at Tianma, China

Variable	Coefficient	Standard Error	t	p>|t|
FEBC area (100 mu)	0.4694***	0.1477	3.18	0.002
Household elevation (100 m)	0.0373	0.0899	0.41	0.678
Household size	0.1715***	0.0543	3.16	0.002
Number of out-migrants	0.1691**	0.0675	2.51	0.013
Number of local off-farm labor	−0.1284	0.1155	−1.11	0.267
Cropland under cultivation (mu)	0.0230	0.0256	0.9	0.369
Gross income (1,000 yuan)	−0.0019	0.0017	−1.09	0.277
Constant	0.9244	0.6539	1.41	0.158

Notes: The model uses data collected from both 2013 and 2014 household surveys with a
sample size of 408 who participated in both GTGP and FEBC.

labor. Thus, FEBC encourages farmers to stay in their original households, farming the croplands with decent quality and/or good accessibility.

6.6 Summary

This chapter examined two concurrent green efforts in TNNR, China: the GTGP
(Policy 1) and the FEBC (Policy 2). We found spillover effects in multiple areas
among the two green efforts; some are synergistic, while others were offset.
Both the GTGP and the FEBC lead to marginal cropland abandonment. The
FEBC tends to make enrolling households continue to rely on fuelwood as the
primary energy source, but the GTGP does not seem to substantially affect fuel

choice or the amount of fuelwood usage. The FEBC may increase enrollment in GTGP (Behavior 1a), which is a positive *Policy–Behavior* spillover effect (Figure 6.6).

On the other hand, FEBC payments (Policy 2) may lead to lower migration rates, an action that GTGP promotes (Behavior 1), constituting a negative *Policy–Behavior* spillover effect. Furthermore, GTGP payments (Policy 1) may also give rise to a higher likelihood of tree theft in FEBC forests (Behavior 2), a *Policy–Behavior* spillover effect that may arise from a *Behavior–Behavior* spillover effect (Figure 6.6). Explicitly referring to the *Behavior–Behavior* spillover effect, migration of the whole family or farm laborers to cities for higher-paying employments (Behavior 1) may lead to a reduction or cessation of monitoring FEBC forests belonging to the household, increasing the chances of timber theft in these forests (Behavior 2). The concurrent PES programs may not maximize their environmental benefits due to the different spillover mechanisms behind household behavior on land-use and livelihood decisions such as labor migration, use of fuelwood, and illegal tree logging. These hidden spillover effects among concurrent green efforts, such as the two PES programs examined in this chapter, have not been recognized in the previous studies. Therefore, future research on PES program evaluations should consider and account for these spillover effects.

References

Barbieri, A. F., Carr, D. L., & Bilsborrow, R. E. (2009). Migration within the frontier: The second generation colonization in the Ecuadorian Amazon. *Population Research and Policy Review*, *28*(3), 291–320. https://doi.org/10.1007/s11113-008-9100-y

Bilsborrow, R. E. (2016). Concepts, definitions and data collection approaches. In M. J. White (Ed.), *International handbook of migration and population distribution* (pp. 109–156). Springer Netherlands. https://doi.org/10.1007/978-94-017-7282-2_7

Bilsborrow, R. E., Barbieri, A. F., & Pan, W. (2004). Changes in population and land use over time in the Ecuadorian Amazon. *Acta Amazonica*, *34*, 635–647. https://doi.org/10.1590/S0044-59672004000400015

Bista, R., Zhang, Q., Parajuli, R., Karki, R., Chhetri, B. B. K., & Song, C. (2021). Cropland abandonment in the community-forestry landscape in the middle hills of Nepal. *Earth Interactions*, *25*(1), 136–150. https://doi.org/10.1175/EI-D-21-0006.1

Chen, X., Zhang, Q., Peterson, M. N., & Song, C. (2019). Feedback effect of crop raiding in payments for ecosystem services. *Ambio*, *48*(7), 732–740. https://doi.org/10.1007/s13280-018-1105-0

Dai, L., Zhao, F., Shao, G., Zhou, L., & Tang, L. (2009). China's classification-based forest management: Procedures, problems, and prospects. *Environmental Management*, *43*(6), 1162–1173.

Leach, G. (1992). The energy transition. *Energy Policy*, *20*(2), 116–123. https://doi.org/10.1016/0301-4215(92)90105-B

Liang, Z. (2016). China's great migration and the prospects of a more integrated society. *Annual Review of Sociology*, *42*(1), 451–471. https://doi.org/10.1146/annurev-soc-081715-074435

Lin, Y., & Yao, S. (2014). Impact of the sloping land conversion program on rural household income: An integrated estimation. *Land Use Policy, 40*, 56–63. https://doi .org/10.1016/j.landusepol.2013.09.005

Ma, X., Heerink, N., Feng, S., & Shi, X. (2015). Farmland tenure in China: Comparing legal, actual and perceived security. *Land Use Policy, 42*, 293–306. https://doi.org/10 .1016/j.landusepol.2014.07.020

Masera, O. R., & Navia, J. (1997). Fuel switching or multiple cooking fuels? Understanding inter-fuel substitution patterns in rural Mexican households. *Biomass and Bioenergy, 12*(5), 347–361. https://doi.org/10.1016/S0961-9534(96)00075-X

Massey, D. S. (1990). Social structure, household strategies, and the cumulative causation of migration. *Population Index, 56*(1), 3–26. https://doi.org/10.2307/3644186

Peng, X. (2011). China's demographic history and future challenges. *Science, 333*(6042), 581–587. https://doi.org/10.1126/science.1209396

Song, C., Bilsborrow, R., Jagger, P., Zhang, Q., Chen, X., & Huang, Q. (2018). Rural household energy use and its determinants in China: How important are influences of payment for ecosystem services vs. Other factors? *Ecological Economics, 145*, 148– 159. https://doi.org/10.1016/j.ecolecon.2017.08.028

Song, C., Zhang, Y., Mei, Y., Liu, H., Zhang, Z., Zhang, Q., Zha, T., Zhang, K., Huang, C., Xu, X., Jagger, P., Chen, X., & Bilsborrow, R. E. (2014). Sustainability of forests created by China's sloping land conversion program: A comparison among three sites in Anhui, Hubei and Shanxi. *Forest Policy and Economics, 38*, 161–167.

Tang, X., & Liao, H. (2014). Energy poverty and solid fuels use in rural China: Analysis based on national population census. *Energy for Sustainable Development, 23*, 122– 129. https://doi.org/10.1016/j.esd.2014.08.006

Uchida, E., Rozelle, S., & Xu, J. (2009). Conservation payments, liquidity constraints and off-farm labor: Impact of the grain for green program on rural households in China. In R. Yin (Ed.), *An integrated assessment of China's ecological restoration programs* (pp. 131–157). Springer Netherlands. https://doi.org/10.1007/978-90-481-2655-2_9

Zhang, L. X., Yang, Z. F., Chen, B., Chen, G. Q., & Zhang, Y. Q. (2009). Temporal and spatial variations of energy consumption in rural China. *Communications in Nonlinear Science and Numerical Simulation, 14*(11), 4022–4031. https://doi.org/10.1016/j.cnsns .2008.04.019

Zhang, Q., Bilsborrow, R. E., Song, C., Tao, S., & Huang, Q. (2018a). Determinants of out-migration in rural China: Effects of payments for ecosystem services. *Population and Environment, 40*(2), 182–203. https://doi.org/10.1007/s11111-018-0307-5

Zhang, Q., Bilsborrow, R., Song, C., Tao, S., & Huang, Q. (2019). Rural household income distribution and inequality in China: Effects of payments for ecosystem services policies and other factors. *Ecological Economics, 160*, 114–127. https://doi.org/10 .1016/j.ecolecon.2019.02.019

Zhang, Q., Hakkenberg, C. R., & Song, C. (2018b). Evaluating the effectiveness of forest conservation policies with multi-temporal remotely sensed imagery: A case study from Tiantangzhai Township, Anhui, China. In S. Liang (Ed.), *Comprehensive remote sensing* (1st ed., Vol. 9, pp. 39–58). Elsevier. https://doi.org/10.1016/B978-0-12 -409548-9.10435-X

Zhang, Q., Song, C., & Chen, X. (2018c). Effects of China's payment for ecosystem services programs on cropland abandonment: A case study in Tiantangzhai Township, Anhui, China. *Land Use Policy, 73*, 239–248.

Zhang, Q., Wang, Y., Tao, S., Bilsborrow, R. E., Qiu, T., Liu, C., Sannigrahi, S., Li, Q., & Song, C. (2020). Divergent socioeconomic-ecological outcomes of China's conversion of cropland to forest program in the subtropical mountainous area and the semi-arid Loess Plateau. *Ecosystem Services*, *45*, 101167. https://doi.org/10.1016/j.ecoser.2020.101167

Zhang, Y., & Song, C. (2006). Impacts of afforestation, deforestation, and reforestation on forest cover in China from 1949 to 2003. *Journal of Forestry*, *104*(7), 383–387. https://doi.org/10.1093/jof/104.7.383

Zhang, Y., Li, X., & Song, W. (2014). Determinants of cropland abandonment at the parcel, household and village levels in mountain areas of China: A multi-level analysis. *Land Use Policy*, *41*, 186–192. https://doi.org/10.1016/j.landusepol.2014.05.011

7   Concurrent green initiatives in Wolong Nature Reserve, China

Hidden spillover effects between the Grain-to-Green Program (GTGP)-like program and the Forest Ecological Benefit Compensation (FEBC) Fund in China are evident in Wolong Nature Reserve in China. Below we present empirical results from a published book chapter by Yang et al. (2016). We first introduce the context setting of Wolong Nature Reserve (also named Wolong National Nature Reserve), then describe data collection and modeling efforts, and finally present findings and implications of the two types of PES programs.

7.1 Wolong National Nature Reserve

Wolong Nature Reserve is located in Wenchuan County, Sichuan Province, in southwestern China, covering a geographic extent of N30°45′–31°25′, W115°42′–115°46′ with a total area of approximately 2,000 km² (Figure 7.1). Wolong was designated in 1975 as a flagship reserve to conserve the endangered giant panda (*Ailuropoda melanoleuca*), a global environmental icon that holds nature's value vital to human society (Liu et al., 2001). Wolong lies in the transition zone between the Qinghai–Tibet Plateau and the Sichuan Basin, ranging from 1,200 to 6,250 m above sea level. The climate is warm and temperate, with a mean annual temperature of 8.9°C and a mean annual rainfall of 995 mm.

Wolong supports a population of 104 giant pandas (Sichuan Forestry Department, 2015), accounting for about 10% of the total number in China. In addition to the giant panda, Wolong is also home to more than 6,000 plant and animal species (He et al., 2008). Over one-third of Wolong's natural landscape is covered by forests, with main vegetation types including evergreen broadleaf, deciduous and coniferous forests, and alpine meadows. These forests provide essential shelter and staple food (e.g., the understory bamboo, *Bashania fangiana*, and *Fargesia robust*) for the wild giant pandas (He et al., 2009) and many other animal and insect species. In 1980, the World Network of Biosphere Reserves under UNESCO's Man and Biosphere Programme recognized Wolong due to its exceptional value for biodiversity conservation. Then in 2006, Wolong was inscribed on the World Heritage List (UNSECO World Heritage Centre, 2006).

DOI: 10.4324/9781003290292-7

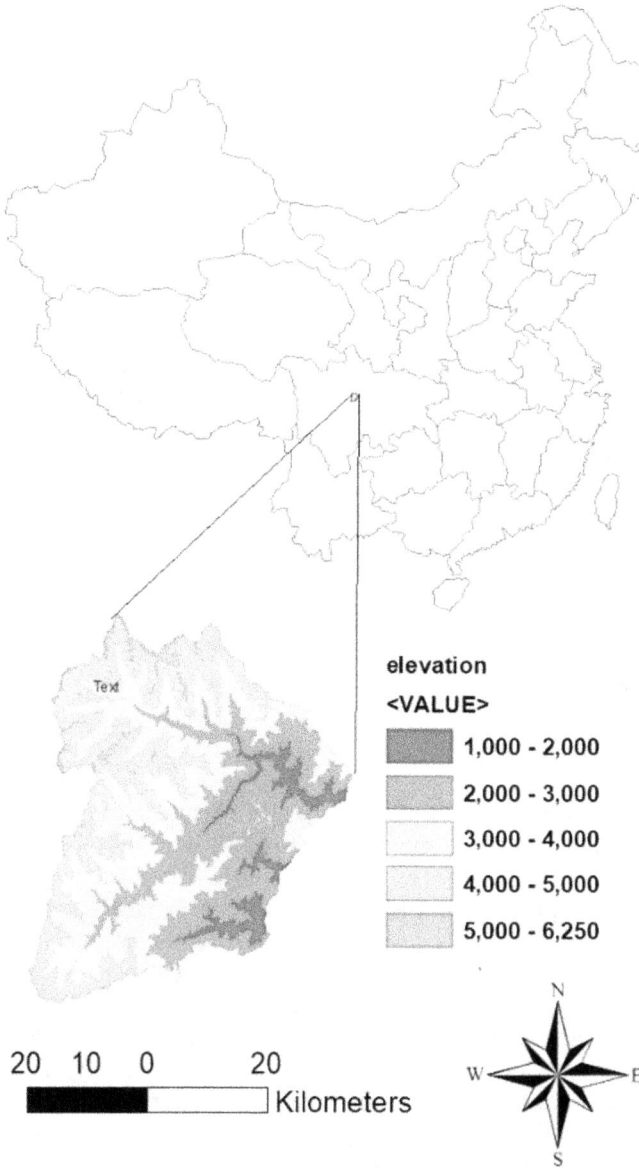

elevation
<VALUE>
1,000 - 2,000
2,000 - 3,000
3,000 - 4,000
4,000 - 5,000
5,000 - 6,250

Figure 7.1 Location of Wolong Nature Reserve, China.

Wolong Nature Reserve, comprised of two townships of Wolong and Gengda, is home to nearly 5,000 rural residents. Most of these residents conduct various subsistence socioeconomic activities such as land cultivation, livestock raising, and fuelwood collection. Local livelihoods rely mostly on forest resources such as fuelwood as the primary energy source for cooking human food and pig fodders and heating during the winter. Thus, these human activities are fundamental drivers of ecosystem degradation, potentially threatening the giant panda in Wolong (Viña et al., 2007).

Around 2001, the reserve implemented the Natural Forest Conservation Program (NFCP) and the Grain-to-Green Program (GTGP) (Yang et al., 2013a). In Wolong, the NFCP annual payment rate ranged from 800 to 1000 yuan per household (or US $97.6–121.9, US $1 = 8.2 yuan in 2001). Households who participated in the GTGP received 240 yuan (US $29.3) per year by converting one mu (1 mu = 1/15 ha) cropland into the forest during the compensation years. With a resemblance to the GTGP, the Wolong government developed and implemented a local PES program named the Grain-to-Bamboo Program (GTBP). The GTBP pays residents cash to convert croplands to bamboo plantations, intending to restore the habitat and provide more stable food for the giant pandas (Yang et al., 2013b). Compared to the GTGP, the GTBP has a much higher annual compensation rate, ranging from 900 to 1200 yuan (or US $109.7–146.3) per mu, as it targets croplands in flatter areas with higher opportunity costs. Since GTGP and GTBP have the same goal and implementation method, they are combined as the Grain-to-Green/Bamboo Program (hereafter referred to as GTGB, 2001–2010) (Yang et al., 2016).

7.2 Data collection

Researchers used panel data collected from several rounds of household interviews from 1999 to 2010 (An et al., 2002; Yang et al., 2016; Yang et al., 2013b), with the first round conducted in the summer of 1999. An et al. first obtained the 1996 Chinese agricultural census list and then adopted the stratified random sampling to draw 220 households (about 23% of the household population) from six villages in the reserve based on the list. They also collected demographic and socioeconomic information through face-to-face interviews (An et al., 2002).

In 2002, 2007, and 2009, researchers conducted household interviews to collect similar information for 200, 192, and 207 households, respectively (Yang et al., 2016; Yang et al., 2013b). For all the four household interview sessions, a total of 179 households were consistently interviewed, making up the final sample for the panel data. Since income was of primary interest, retrospective data on household income and expenditure were collected every year from 1998 to 2009 during the household interviews. In the 2007 and 2009 surveys, supplemental questions about the three PES programs (i.e., NFCP, GTGP, and GTBP) included asking for information about payments received by each surveyed household and their perception of the programs. All monetary measures, such as income and expenditure, were deflated according to the 2000 consumer price index (Yang et al., 2016).

7.3 Data analysis and modeling

According to our definition, Yang et al. investigated economic transformation among local households in Wolong under GTGB and NFCP, two concurrent programs. Individually, Yang et al. examined the association between household income growth from 1998 to 2007 and household payments (Yang et al., 2016). The hypothesis sought to explain income growth through relationships to factors representing macro-socioeconomic conditions and household-level characteristics. Macro-socioeconomic conditions refer to the effects of the two PES programs and other policies. Household-level factors were represented by the household's access to different forms of capital, including financial, human, natural, built-up, and social capitals. The dependent variable and explanatory variables are summarized in Table 7.1.

The general regression model can be written as:

$$Y = \alpha + \beta_1 P_1 + \beta_2 P_2 + \gamma C + \delta D + \varepsilon \qquad (7.1)$$

where Y is a vector of household income growth from 1998 to 2007, calculated by subtracting total income in 1998 from total income in 2007. P_1 is a vector of policy intervention, represented as the amount (or percentage) of annual payment received by, or the participation status of, each household in the programs. P_2 is a vector of interactions between the programs. C is a vector of variables reflecting households' access to the five forms of capitals (Table 7.1). D is a contextual variable controlling for regional differences between the two townships (Wolong and Gengda). Regarding the parameters, α is the intercept, while β_1, β_2, γ, and δ capture the fixed effects of the vectors of programs, interactions of the programs, capital access, and contextual factor, respectively. Finally, ε is the error term assumed to be normally distributed with a mean of zero and constant variance.

7.4 Findings and discussion

Although not directly addressing potential spillover effects between the two concurrent PES programs, the regression results (Table 7.2) provide implicit evidence of their interacting effects in Wolong. We found that the GTGB (or the NFCP) payment as part of the total household income had a statistically significant negative impact on household income growth from 1998 to 2007 when considering the program alone controlling other factors. For instance, with each 1% increase in NFCP income, household income growth from 1998 to 2007 decreased by 1280 yuan when controlling for other factors. Similarly, a 1% increase in GTGB payment percentage in total income resulted in a decrease in income growth by 150 yuan.

Surprisingly, the interaction of NFCP and GTGB (i.e., GTGB payment percentage in total income × NFCP payment percentage in total income) had a positive impact on the above income difference (Yang et al., 2016). To better understand

Table 7.1 Description of income growth and explanatory variables at Wolong, China

Variable	Description	Mean	Std. dev.
Dependent variable			
Income growth	Difference in total household income in 2007 subtracting total household income in 1998 (1,000 yuan)	21.988	27.286
Policy variable			
NFCP payment	Amount of annual payment received from NFCP (1,000 yuan)	0.948	0.183
NFCP percentage	Percentage of annual payment received from NFCP in total household income in 2007	6.5%	6.0%
GTGB payment	Amount of annual payment received from GTGB (1,000 yuan)	2.888	2.320
GTGB percentage	Percentage of annual payment received from GTGB in total household income in 2007	16.0%	15.3%
ESP subsidy	Amount of initial subsidy for electricity consumption received from ESP (1,000 yuan)	0.086	0.104
ESP percentage	Percentage of initial subsidy for electricity consumption received from ESP (1,000 yuan)	2.7%	4.1%
TDP participation	Household participation status in tourism business (1 = participated; 0 = did not participate)	0.274	0.447
Financial capital			
Initial total income	Total household income in 1998 (1,000 yuan)	6.285	4.932
Initial percentage of agricultural income	Percentage of agricultural income in total household income in 1998	63.0%	31.3%
Change in agricultural income	Difference in household agricultural income in 2007 subtracting household agricultural income in 1998 (1,000 yuan)	7.817	15.393
Human capital			
Number of labor	Number of laborers in household	2.820	1.455
Change in number of laborers	Difference in the number of household laborers in 2007 subtracting the number of household laborers in 1998	−0.727	1.795
Education	Education level of the most educated non-student adult in 2007 (year)	7.120	3.432
Natural capital			
Cropland area	Total area of cropland in 2007 (mu, 1 mu = 1/15ha)	10.450	4.163
Built-up capital			
Distance to the main road	Euclidean distance from household location to the main road (km)	0.431	0.629
Social capital			
Social ties to local governments	Whether the household had an immediate relative member working in local governments or government enterprises (1 = yes, 0 = no)	0.120	0.326

Notes: NFCP: Natural Forest Conservation Program, i.e., one component of the Forest Ecological Benefit Compensation (FEBC) Fund; GTGB: Grain-to-Green/Bamboo Program; ESP: Electricity Subsidy Program; TDP: Tourism Development Program; N/A: not available. The results are from Yang et al. (2016).

Table 7.2 Regression results from Wolong, China

Variable	Coefficient	Robust SE
Policy variables		
NFCP percentage	−128.811***	35.194
GTGB percentage	−15.535**	7.707
ESP percentage	63.921**	28.004
TDP participation	5.274*	3.188
NFCP percentage × GTGB percentage	387.458***	89.473
NFCP percentage × ESP percentage	−188.653	133.730
NFCP percentage × TDP participation	−228.758**	98.906
Financial capital		
Initial total income	−0.175	0.294
Change in agricultural income	1.114***	0.251
Human capital		
Number of labor	2.767**	1.283
Change in number of labor	2.161**	0.977
Education	0.119	0.407
Natural capital		
Cropland area	−1.451**	0.578
Built-up capital		
Distance to main road	−1.783***	0.563
Social capital		
Social ties to local governments	0.319	3.919
Contextual factor		
Township	4.253	2.671
Constant	14.601***	2.788

Notes: *$p<0.10$; **$p<0.05$; ***$p<0.01$. The regression results stand for the effects of concurrent PES programs (NFCP and GTGB) and other control variables on changes in total household income from 1998 to 2007 at Wolong, China (Yang et al., 2016).

the interaction effects, we use an example to illustrate how one PES program may affect income growth through the other PES program. Given a household with a mean value of GTGB payment percentage in total household income (0.16), the conditional effect of 1 unit (i.e., 1%) of NFCP payment percentage on income growth depending on the value of GTGB payment percentage is [(−128.8 + 387.5 × 0.16)/100] = −0.6681772 (unit: thousand yuan). With every additional 1% NFCP payment percentage in total income, combined with the 16% GTGB payment percentage in total income, it would cause overall income growth to decline by 668.1772 yuan.

The conditional effects of NFCP (or GTGB) payment share in total income are affected by GTGB (or NFCP) payment shares of total income. Based on the plots (Figure 7.2), the conditional NFCP effect remains negative when the GTGB payment share of total income is below 0.33, but it becomes neutral (or positive) when the GTGB payment share reaches (or exceeds) the threshold. Simultaneously, the conditional GTGB effect, when interacting with the NFCP effect, would flip from negative to positive if the NFCP payment share in total income increases beyond 0.04.

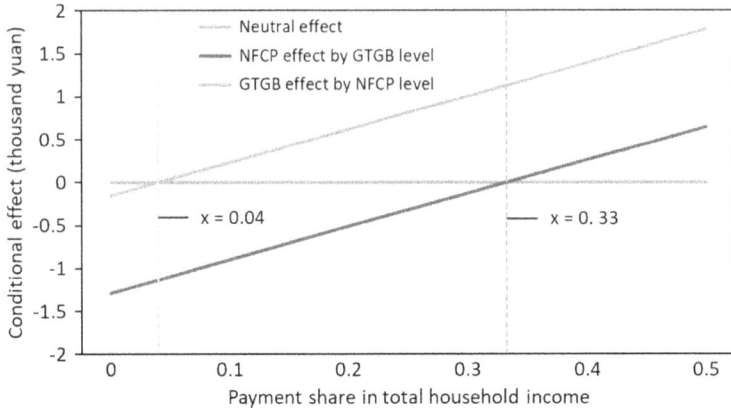

Figure 7.2 Combined effects of NFCP and GTGB on household income difference by different NFCP or GTGB payment shares of total income. The right dash line denotes the threshold (0.33) of GTGB payment share in total income where the total NFCP effect becomes zero; the left dash line denotes the threshold (0.04) of NFCP payment share in total income where the total GTGB effect becomes zero.

The derived effects under the scenarios of single policy and concurrent policies represent the hidden linkages due to the simultaneously implemented programs. Each of the two PES programs alone has a negative effect on income growth, but their synergistic effect is positive, more robust than each individual effect. This surprising outcome may result from households' livelihood changes in adaptation to the implementation of the policy. Under the scenario of GTBP only, "transformative" changes are likely to happen as households shift their livelihoods from farming work to non-agricultural activities, such as rural-to-urban migration, featuring *Policy–Behavior* and *Behavior–Behavior* spillover effects. Thus, households seek non-agricultural opportunities, bringing an economic return to compensate for the reduced income from farming land. Under the NFCP only, household livelihoods tend to experience "incremental" changes, intensifying agriculture by growing more cash crops and using more fertilizer or pesticides. This change characterizes an internal *Policy–Behavior* effect. Compared to alternative livelihood activities such as local off-farm work and migration, the agricultural intensification may contribute less to the total household income growth in the long term since agricultural outputs (e.g., crops) are less lucrative for income but primarily for self-consumption.

The ecosystem outcomes under Behaviors 1 and 2 were not directly addressed in Yang et al.'s (2013) work, and there is evidence elsewhere showing both are positive (Viña et al., 2007). Goals 1 and 2 are similar since both refer to enhanced forest cover and panda habitat quality, except that the latter should be more significant as local villagers switched to non-agricultural activities. Considering the

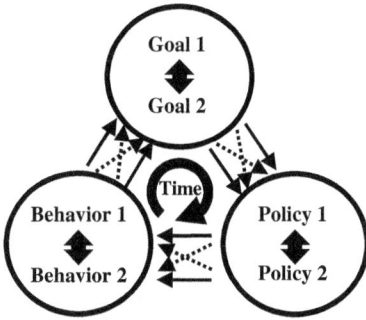

Figure 7.3 Cross-program spillover effects at Wolong, China. The diagram is modified from Figure 1.3, where the solid one-way arrows stand for internal influences from one element to another within the same initiative while the dashed one-way arrows and double two-way arrows for potential spillover effects; the circular one-way arrow represents Time–Time spillover effects. The shaded arrow represents the spillover effect with evidence from this section.

Goal–Policy spillover effect, how changes in ecosystems may feedback into the concurrent payments of NFCP and GTGB is unclear and needs more in-depth analysis. Furthermore, NFCP payments were renewed in 2008 for another 10 years, and GTGB payments ceased in 2010, which complicated the policy effects and made uncovering the hidden linkages more needed.

7.5 Summary

GTGB and NFCP, when implemented alone, each impeded the growth of household income from 1998 to 2007. However, they positively impacted this income growth when applying to the two payments together (Figure 7.3). This surprising outcome is likely due to changes in livelihood strategies: local farmers switched from agricultural intensification (Behavior 1) to out-migration (Behavior 2), substantially reducing or even abandoning farming activities. This *Behavior–Behavior* spillover effect may not be apparent when evaluating each PES program in isolation.

References

An, L., Lupi, F., Liu, J., Linderman, M. A., & Huang, J. (2002). Modeling the choice to switch from fuelwood to electricity: Implications for giant panda habitat conservation. *Ecological Economics*, *42*(3), 445–457.

He, G., Chen, X., Beaer, S., Colunga, M., Mertig, A., An, L., Zhou, S., Linderman, M., Ouyang, Z., Gage, S., Li, S., & Liu, J. (2009). Spatial and temporal patterns of fuelwood collection in Wolong Nature Reserve: Implications for panda conservation. *Landscape and Urban Planning*, *92*(1), 1–9.

He, G., Chen, X., Liu, W., Bearer, S., Zhou, S., Cheng, L. Y., Zhang, H., Ouyang, Z., & Liu, J. (2008). Distribution of economic benefits from ecotourism: A case study of

Wolong Nature Reserve for Giant Pandas in China. *Environmental Management, 24*(6), 1017.

Liu, J., Linderman, M., Ouyang, Z., An, L., Yang, J., & Zhang, H. (2001). Ecological degradation in protected areas: The case of Wolong nature reserve for giant pandas. *Science, 292*(5514), 98–101.

Sichuan Forestry Department. (2015). *The fourth survey report on giant panda of Sichuan.* Sichuan Science and Technology Press.

UNSECO World Heritage Centre. (2006). *Sichuan giant panda sanctuaries.* https://whc .unesco.org/en/list/1213/

Viña, A., Bearer, S., Chen, X., He, G., Linderman, M., An, L., Zhang, H., Ouyang, Z., & Liu, J. (2007). Temporal changes in connectivity of giant panda habitat across the boundaries of Wolong, China. *Ecological Applications, 17*(4), 1019–1030.

Yang, W., Dietz, T., Liu, W., Luo, J., & Liu, J. (2013a). Going beyond the Millennium Ecosystem Assessment: An index system of human dependence on ecosystem services. *PloS One, 8*(5), e64581.

Yang, W., Liu, W., Viña, A., Luo, J., He, G., Ouyang, Z., Zhang, H., & Liu, J. (2013b). Performance and prospects of payments for ecosystem services programs: Evidence from China. *Journal of Environmental Management, 127*, 86–95. https://doi.org/10 .1016/j.jenvman.2013.04.019

Yang, W., Lupi, F., Dietz, T., & Liu, J. (Jack). (2016). Dynamics of economic transformation. In J. Liu, V. Hull, W. Yang, A. Viña, X. Chen, Z. Ouyang, H. Zhang. (Eds.), *Pandas and people: Coupling human and natural systems for sustainability* (pp. 109–119). Oxford University Press.

8 Spillover effects worldwide

From here on, we concentrate on evidence of the spillover effects based on nine cases outside of China and the USA. Rather than presenting on a case basis, we summarize what we found based on our conceptual framework (Figure 1.3). To cover all the spillover effects, we still mention the findings from the USA and China cases but refer to Chapters 5–7 for details.

8.1 *Policy–Behavior* cross-program spillover effects

Policy–Behavior spillover effects refer to one (kind of) policy that may change actions expected from another policy (Figure 1.3). To illustrate such spillover effects, we first examined two concurrent payments in China. China launched its Grain-to-Green Program (GTGP) from 1999 to 2001, providing farmers with cash and/or grain subsidies to convert croplands on steep slopes or otherwise ecologically sensitive areas into forestland or grassland. In 2001, China started another program providing extensive payments for environmental services (PES)—the Forest Ecological Benefit Compensation (FEBC) Fund—seeking to protect and manage natural forests for public benefit. Since 2004, these two programs have been implemented simultaneously in 20 provinces, autonomous regions, and municipalities. GTGP-eligible land parcels are farmland on steep slopes, whereas FEBC parcels are natural forestlands, thus spatially disconnected from GTGP parcels. In many regions, parcels of both types of land are contracted to the same households (Yost et al., 2020), making them horizontally stacked payments.

Our case studies show that *Policy–Behavior* spillover effects have occurred in two nature reserves in China. Using data from Fanjingshan National Nature Reserve, China, we found that each 1,000 Yuan of FEBC payment (Policy 2) would increase land enrollment (Behavior 1) in response to GTGP payment (Policy 1) by 0.4487 mu (1 mu = 1/15 ha; $p = 0.0064$; Table 5.2). Compared to GTGP, the payment rate of FEBC is much lower while the land area of FEBC is much higher (Table 4.1); households in Tianma National Nature Reserve, China, received about twice as much compensation from FEBC as that from GTGP. Similar *Policy–Behavior* spillover effects are found in Tianma: under FEBC payment (Policy 2), every 100 mu of forest land generates 0.47 mu more land enrolled in GTGP (Policy 1, $p = 0.0020$; Table 6.7) based on our regression model

DOI: 10.4324/9781003290292-8

when controlling for other confounding variables (for the potential mechanism, see Section 6.5.5).

Policy–Behavior spillover effects have also taken different forms or pathways. Every additional 1,000 yuan of FEBC payment (Policy 2) decreases the odds of out-migration (Behavior 1) by 34% (Table 6.3) at Tianma. However, the classic PES literature has long indicated a *Policy–Behavior* internal link: payment from GTGP (Policy 1) may trigger and/or facilitate out-migration (part of Behavior 1), which is also observed by our data for the Tianma case. Also, at Tianma, every 100 m closer to the nearest GTGP (Policy 1) land increases the odds of FEBC tree theft (Behavior 2) by 15.5% (Section 6.5.4), suggesting that FEBC trees are more likely to be illegally logged by other rural residents if a household is located in proximity to GTGP land. For the potential mechanism, see the analysis regarding hidden *Behavior–Behavior* spillover effects in Section 8.5.

8.2 *Behavior–Goal* spillover effects

Behavior–Goal spillover effects refer to the situation where behaviors expected from one policy might give rise to unexpected changes in the goal(s) targeted by another policy (Figure 1.3), illustrated in our case for Australia. The country has 85.3 million ha of intensive agricultural land, subject to reforestation under carbon farming policies. Modeling by Bryan et al. (2015) shows that under a policy scheme that focuses on carbon sequestration, people would establish "carbon plantings" (Behavior 1) of fast-growing Eucalyptus monocultures (Behavior 1), which sequesters a large amount of carbon (i.e., a strong internal *Behavior–Goal* link) but adds little to biodiversity (i.e., a weak *Behavior–Goal* spillover effect) (Bryan et al., 2015). Under a policy scheme that highlights both carbon and biodiversity services, the practice of "environmental plantings" (Behavior 2; mix of native trees and shrubs) gives rise to not only high levels of carbon sequestration (a strong internal *Behavior–Goal* link; only 1.32% of total carbon stock sacrificed) but also a significant gain in biodiversity (a strong *Behavior–Goal* spillover effect)—96 times that from the carbon plantings (Bryan et al., 2015).

Behavior–Goal spillover effects are also evidenced at the Mazar Wildlife Reserve in Ecuador (Bremer et al., 2016). PROFAFOR (Programa FACE de Forestación del Ecuador; Policy 1) aims to promote afforestation with *Pinus* species and some native Andean species (Behavior 1) in the hope of enhancing carbon sequestration (Goal 1). At the same time, the SocioPáramo program (a subprogram of the more extensive SocioBosque program; Policy 2) seeks to exclude burning in Páramo grasslands (Behavior 2) for multiple ecosystem services of carbon storage, biodiversity protection, and water provision (Goal 2). However, afforestation (Behavior 1) has caused decreased soil moisture and loss of native plant diversity, compromising Goal 2; the soil was significantly drier under pines, having a volumetric soil moisture content of 13–22% compared to 50–74% at grassland sites. In pine plantation plots, the estimated species richness decreased to 42% of the native Páramo communities. In the same site, the burning-exclusion action (Behavior 2) may not achieve optimal carbon sequestration results (Goal 1), representing another *Behavior–Goal* spillover effect.

8.3 *Goal–Policy* spillover effects

Goal–Policy spillover effects imply that goals or outcomes (e.g., Goal 1) generated or enhanced from one policy (Policy 1) may loop back to affect the other policy(s) (Policy 2; Figure 1.3). Recently in the USA, a concurrent PES scheme called PES stacking has emerged. However, "there are no regulations addressing stacking or any guidance documents from US federal resource agencies" (Robertson et al., 2014), nor any evidence-based guidelines about how to achieve or improve the intended ecosystem services.

Evidence of *Goal–Policy* spillover effects was found in the Neuse River Basin in North Carolina, USA (Program Evaluation Division, 2009). The North Carolina Department of Transportation paid US$3.5 million for wetland credits in 2000, aiming to restore ecosystem services on 438.5 acres of wetlands. Of these 438.5 acres, 69.5 acres were used by another government agency—the Division of Water Quality (under the North Carolina Department of Environment and Natural Resources)—to certify nutrient offset credits in 2008. Of these 69.5 certified acres, 46 acres received US$698,372 for nutrient offset credits in 2009. This payment of US$698,372, temporally stacked on the same 46 acres that had received wetland payment, was considered "double-dipping" for generating no additional value. In response to this controversy and related public pressure, the North Carolina Division of Water Quality decided to rescind this stacking in the future while still honoring the overlapping nutrient offset credits certified in 2008. Assuming a uniform payment rate for the 438.5 acres of wetland, the 46 acres of wetland should have received US$367,160 for wetland credits in 2000. Given that the same 46 acres of wetland had also received a payment of US$698,372 for nutrient credits in 2008, such stacking of payments has amounted to a "double-dipping" rate of 190% (i.e., US$698,372/US$367,160), implying a big waste in conservation payments.

8.4 *Policy–Policy* spillover effects

Policy–Policy spillover effects occur when one policy directly leads to changes in another policy (Figure 1.3), as in the Baltic Sea case (Gren & Elofsson, 2017). To counter the severe eutrophication problem in the Baltic Sea, nine countries in the catchment (i.e., Denmark, Finland, Germany, Poland, Sweden, Estonia, Latvia, Lithuania, and Russia) have agreed to reduce nitrogen (N) and phosphorous (P) loads entering the sea. Specifically, these countries implemented several abatement measures that reduce the total N and P loads below predetermined annual limits. To minimize total abatement costs, N and P emission permits can be traded in markets among various actors (e.g., abatement firms), between upstream and downstream areas, and across different abatement measures as long as the total N and P caps are observed (Gren & Elofsson, 2017). Gren and Elofsson have demonstrated mathematically that, to be cost effective, payments for N and P abatement (Policies 1 and 2) must coexist and be stacked (Gren & Elofsson, 2017). Many abatement actions, such as wetland construction, generate N and P reductions. If only one policy (through markets of N or P credits) is allowed, the outcome would be much worse, e.g., either higher costs or caps not observed (Gren & Elofsson, 2017).

The Jordan Lake case provides additional evidence of *Policy–Policy* spillover effects, where the stacking of two payments may or may not function well depending on the relative sizes of the payments (Motallebi et al., 2018). In this case, the primary goal—ecosystem service—is the reduction of N loads into Jordan Lake (Goal 1) by all farmers in the watershed, as required by the water quality trading program in North Carolina. In a hypothetical scenario, Goal 2 is considered to reduce P by providing P credits. Both goals are a joint outcome of single conservation practice: building or extending a vegetated riparian buffer (Behavior 1), which aims to increase Goal 1. When the amount of payments under Policy 2 (for P reduction) is between 20% and 30% of Policy 1 (for N reduction), the so-called "double-dipping" occurs, in which a stacked payment (under Policy 2) increases farmers' revenue but does not change their conservation action (Behavior 1).

Complementary to the Jordan Lake case, the Rio Grande catchment case from Bolivia presents spillover effects between concurrent payments made to different parcels but contracted to the same individuals (i.e., horizontal stacking) to conserve biodiversity and improve water quality (Goal 1) (Bottazzi et al., 2018). Payments at level 1 (Policy 1), which are much higher in amount and stricter in monitoring for compliance, seem to downgrade or nullify payments made at levels 2 and 3 targeted on different lands (Policy 2). In this case, local farmers were less compliant with their contracts that required them to stop farming (Behavior 2).

8.5 *Behavior–Behavior* cross-program spillover effects

Behavior–Behavior spillover effects refer to cross-behavior influences, where payment-induced actions affect each other. Our first evidence comes from Wolong Nature Reserve, which is conserved for the giant panda (*Ailuropoda melanoleuca*) and many other plants and animals of high value in the same area. Two concurrent payments exist: The National Forest Conservation Program (NFCP) and a GTGP-like program called Grain to Bamboo Program (GTBP). Yang et al. (2016) found that payments from GTBP and NFCP, when implemented alone, each had a significant negative impact on the growth in household income from 1998 to 2007. However, the two payments positively impacted this income growth when implemented together. This surprising outcome could arise from local people's changes in their livelihood strategies; local farmers switched from agricultural intensification (Behavior 1) to out-migration (Behavior 2), substantially reducing or even abandoning farming activities (detail in Chapter 7).

The Australian case also supports this *Behavior–Behavior* type of spillover effects. The two actions, i.e., establishment of "carbon plantings" (CP; Behavior 1) and "environmental plantings" (EP; Behavior 2), are subjected to a quantitative restriction; the sum of the CP area, the EP area, and the traditional cropland area should be 85.3 million ha, where increases in CP are coupled with decreases in EP due to constraints in total budget and area of land available (Bryan et al., 2015). *Behavior–Behavior* spillover effects can also be observed at Tianma National Nature Reserve in Anhui Province, China. As shown in our discussion of *Policy–Behavior* spillover effects (Section 8.1), payments from GTGP (Policy 1) may lead to an unintended behavior of tree theft on FEBC land (Behavior 2). This *Policy–Behavior* spillover effect may arise from a *Behavior–Behavior* spillover effect: migration of the whole family or farm laborers to cities for higher-paying employment (Behavior 1) may fail to monitor FEBC forests belonging to the corresponding households, increasing the chances of timber theft in these forests (Behavior 2).

8.6 *Goal–Goal* spillover effects

Goal–Goal spillover effects arise when ecosystem services, especially those implemented nearby or in a co-located manner, interact with one another via various biophysical and ecological processes operating at various spatial and temporal scales (Figure 1.3). Here we use the Foglia River Basin and Marecchia River Basin in Italy as an example. Among the forest-based ecosystem services identified by Morri et al. (2014), water retention (Goal 1) and drinking water supply (Goal 2) are linked conceptually and quantitatively; water retention—a function of forest type (which determines the percentage of runoff retained) and its area—is the source of drinking water. In addition, the two goals of soil protection (amount of soil erosion avoided) and CO_2 sequestration are both a function of forest type and its area, with a few other variables under control. For the policy schemes related to the above goals and the potential "double-dipping" problem, we refer to Sections 2.3 and 8.3.

Data from the New World (the Americas and Oceania) and Great Britain may contribute to understanding these *Goal–Goal* spillover effects. Compared to a carbon-only (Goal 1) strategy, a combined carbon-biodiversity strategy—weighing the two goals and adjusting subsequent spatial allocation—could simultaneously protect 90% of carbon stocks and more than 90% of the biodiversity protected under a biodiversity-only (Goal 2) strategy. This win–win gain arises from heterogeneous spatial distributions of—and site-specific interactions between—biodiversity and carbon goals (Thomas et al., 2013). Similar win–win outcomes are also observed in the Australia case due to reallocating payments to sites with abundant biodiversity and carbon goals (Bryan et al., 2015).

Goal–Goal spillover effects are also found in the Neuse case, where Goal 1 (derived from wetland payment) automatically entails Goal 2 (derived from nutrient payment). This *Goal–Goal* spillover effect becomes the rationale for the aforementioned *Goal–Policy* spillover effect at Neuse: Goal 1 (wetland credits) already existed and should continue generating co-benefits of nutrient offset (Goal 1) for which Policy 2 (nutrient offset credits) is intended, leading to the North Carolina

Division of Water Quality's decision to rescind Policy 2. Similarly, in the Jordan Lake case, nitrogen reduction (Goal 1) that comes with constructing riparian buffers would bring in phosphorous reduction (Goal 2) because of N and P cycling processes (Motallebi et al., 2018). In the context of many ecological services of GTGP and FEBC, it has been reported that such *Goal–Goal* spillover effects exist. Once forests are established under GTGP that are often closer to households, FEBC forests are better protected, as GTGP forests may act as buffers for human activities such as fuelwood collection and grazing that would otherwise occur in FEBC forests (Song et al., 2018).

8.7 Yucatán and Chiapas

Mexico is a pioneer Latin American country in implementing a nationwide Ecosystem Services-Hydrological program (PSA-H) policy to protect critical forests for water provision and regulation services. Beginning in 2003, the PSA-H payments were granted to forest communities for five consecutive years after signing a contract by community elected leaders and Mexico's National Forestry Commission (CONAFOR). Simultaneously, another concurrent PES program was named the Forest Ecosystems Conservation and Restoration Program (PROCOREF in Spanish). According to Ezzine-de-Blas et al.'s survey of 77 communities (*ejidos*) in 2013 in Southern Yucatán (Ezzine-de-Blas et al., 2016), we performed a correlation analysis among the area (unit: ha) of enrollment in PSA-H (ha_psa), the payment that each community received from PSA-H in 2012 (value_psa), and the area (unit: ha) of enrollment in PROCOREF (procoref1_ ha). The goal is to examine whether spillover effects exist between the two PES programs.

PSA-H had a positive correlation with the area of PROCOREF enrollment in terms of both area ($r=0.3453$, $p=0.1360$) and amount of payment ($r=0.3500$, $p=0.1303$), suggesting potential *Behavior–Behavior* spillover effects (Table 8.1). However, these two relationships were insignificant (but close to significant) at the alpha$=0.10$ level (Table 8.1). The effective sample size was only 20 compared to a total sample size of 77. As the original paper by Ezzine-de-Blas (2016) did not focus on cross-program spillover effects, most records lacked data in one, two, or all of the three payment-related variables. This case shows that a lack of explicit focus on cross-program spillover effects would limit the usefulness of such data when examining cross-program spillover effects.

8.8 *Time–Time* spillover effects

Time–Time spillover effects occur among various payments that evolve (Figure 1.3), depending on the socioecological context in which they are embedded. As shown previously, we have observed positive spillover effects from FEBC payment to GTGP enrollment in China's two nature reserves. To explore whether existing spillover effects may change over time, we designed questions in a household survey, asking local villagers about their willingness to enroll

Table 8.1 Results of Pearson correlation analysis

Variable	Area of enrollment in PSA-H (unit: ha)	Payment that each community received from PSA-H in 2012	Area of enrollment in PROCOREF (unit: ha)
Area of enrollment in PSA-H (unit: ha)	1.0000 51	0.9756 (<0.0001) 51	0.3453 (0.1360) 20
Payment that each community received from PSA-H in 2012	0.9756 (<0.0001) 51	1.000 51	0.3500 (0.1303) 20
Area of enrollment in PROCOREF (unit: ha)	0.3453 (0.1360) 20	0.3500 (0.1303) 20	1.0000 28

Note: Correlation coefficient, *p*-value, and number of records used in the analysis (sample size) are included in the table.

more cropland in GTGP under a set of hypothetical conditions. Our data analysis revealed a negative relationship: each mu of FEBC land (a proxy of FEBC payment) decreased the odds of GTGP participation by 0.30% (equivalent to a probability decline of 3.0% for a typical household at Fanjingshan). This finding is also corroborated by a similar study conducted one year earlier at the same site (Yost et al., 2020), which found that each mu of the FEBC land would decrease the odds of GTGP participation by 0.58%. Although these negative spillover effects are minor in magnitude, they have evolved from significant positive spillover effects (Table 5.4) over a relatively short period. Farmers with more FEBC land may have enrolled most of their eligible cropland parcels in GTGP, leaving little additional land for the hypothetical GTGP. Aside from this land-scarcity issue, concerns for food security may preclude farmers from enrolling additional cropland (Yost et al., 2020).

The PVPF-KPWS case in Cambodia also shows evidence of *Time–Time* spillover effects (Clements et al., 2010). Payments from the Bird Nest Protection (BNP) program (Policy 1) are made to eligible individuals who then locate, monitor, and protect the remaining nesting sites (Behavior 1a). However, villages receiving such BNP payments allowed in-migrants to settle locally (Behavior 1b); these in-migrants tended to clear forests and cause a more significant loss of bird habitat, offsetting the conservation effects of BNP in the long run. On the other hand, payments from the Ecotourism and Agri-Environment (E&AE) programs (Policy 2) may take several years to build up the capacity of all participating villages and individuals. However, once such capacity is established, such payments may lead to restraining in-migration (Behavior 1b) and the associated deforestation, contributing to bird conservation. The best conservation outcome may come from sequential implementations of these two payments: the BNP first (for immediate effect) and E&AE later (for long-term protection), manifesting a *Time–Time* spillover effect.

8.9 Intertwined spillover effects

As earlier sections show, spillover effects could co-occur, manifesting in an inter-twined, multi-dimensional style. We examine two concurrent green initiatives that are widely implemented globally as an example. The first one is Community-based Forest Management (CFM), which currently manages about 18% of the global forest in 62 countries, supporting hundreds of millions of people (Gilmour, 2016). The CFM is decentralized management of natural resources, aiming at sustainable forest development and livelihood improvement for the local com-munities participating in forest management. CFM devolves the right of forest management decision-making to the local communities and the responsibility of forest conservation. Devolvement of responsibility from the central government to the local communities is perhaps the most significant paradigm shift in forest management policy since the mid-1980s when the central government dominated forest management decision-making without considering the need of the local communities living around the forests. The centralized forest management has led to widespread deforestation and forest degradation, such as the well-known Himalaya Ecological Crisis (Eckholm, 1975).

The second green initiative is the Reducing Emissions from Deforestation and forest Degradation, plus sustainable forest management, conservation, and enhancement of forest carbon stocks (REDD+), which was developed by the Parties to the UN Framework Convention on Climate Change. REDD+ provides results-based payments for the carbon stored in forests in developing countries. REDD+ has been recognized as an effective mechanism for global warming miti-gation. As of 2019, UN-REDD had made significant progress toward REDD+ par-ticipating countries, primarily in the developing world, toward REDD+ goals (http://www.un-redd.org). For any developing country with rich forest resources to receive payments from REDD+ for carbon storage, it first has to develop a national policy, then implement the policy, and finally fully measure, report, and verify the implementation results. Because CFM has improved forest conditions worldwide, many REDD+ pilot projects have been implemented in community forests. Dynamic spillover effects occur between the two policies and the subse-quent actions and gains.

REDD+ and CFM can mutually benefit from each other because of the shared goal of sustainable forest management. The REDD+ payment for carbon storage can benefit the members of Community Forest User Groups (CFUGs). At the same time, REDD+ can take advantage of the natural, social, and institutional capitals that have been accumulated through CFM to achieve its goals (Newton et al., 2015). The improved forest conditions under CFM provide the biophysical basis for REDD+ projects. The bonding social capital developed over time within the CFUGs and the institutions for forest governance would significantly benefit REDD+ project management. However, there are also divergent goals between the two initiatives. Conservation and enhancement of forest carbon stocks for global warming mitigation pursued by REDD+ may compromise the use of for-est products in community forests for livelihood support for forest-dependent people under CFM. Although REDD+ payments are made to the corresponding forest management communities for carbon storage, the payments may or may not make up for the loss of support they used to derive from the community forests (Maraseni et al., 2014; Marquardt et al., 2016). The use of forest products often constitutes carbon leakage from the community forests, compromising the REDD+ goals.

Key actors in formulating REDD+ policy include the national government, international donors, NGOs, and civil society organizations, while the local communities where the REDD+ projects are implemented were generally not well engaged or informed (Bastakoti & Davidsen, 2014). Such a top-down approach leads to the recentralization of forest management, i.e., the local community loses the autonomy for forest management decision-making to utilize the forest resources for livelihood support, such as generating necessary revenues for local community needs and poverty alleviation through timber trade (Phelps et al., 2010). However, CFM improved forest conditions under its management regime. Not all CFUG activities contribute to conserving and enhancing forest carbon storage. Therefore, implementing the REDD+ project in a community forest leads to changes in its governance rules in CFM, which could create opportunities for elite capture of benefits while restricting the poor and marginalized people's use of forest resources (Poudel et al., 2014). Poor people who depend more on forest resources for livelihoods would be disproportionally impacted as a result (Devkota & Mustalahti, 2018). In some cases, the implementation of REDD+ seemingly enhanced the participation in decision bodies by the poor, women, and marginalized groups, but the benefits of REDD+ did not trickle down to these people (Devkota, 2020).

The primary goal of REDD+ projects is global warming mitigation via carbon removal from the atmosphere through forest growth. Although CFUGs have use rights to timber, firewood, and fodder in the community forests, there is no clear legal ownership right for the carbon accumulated in community forests. Clarification of benefit distribution of REDD+ payments becomes critical for its success. Verifiable forest carbon storage is the only criterion for receiving monetary compensation from REDD+. Although REDD+ supports sustainable forest management, no other sustainable forest management metrics beyond carbon storage receive REDD+ compensation.

In contrast, the benefits generated in community forests are multi-dimensional, including forest conservation, sustainable forest management, support for livelihoods for the local people, preservation of biodiversity, and carbon storage for global warming mitigation that benefits the entire world. The commodification of carbon via REDD+ projects could overrun community forest priorities (Bastakoti & Davidsen, 2014). The REDD+ payments to poor households are sometimes insufficient for livelihood enhancement activities (Shrestha et al., 2017).

REDD+ projects can be successfully implemented in forests under CFM (Sharma et al., 2020). However, the multiple levels of spillover effects between REDD+ and CFM have to be addressed appropriately in REDD+ policies and the process of REDD+ implementation, including a clear definition of carbon ownership and benefit distribution mechanism, carbon price, preservation of CFUGs' autonomy in forest management decision-making, and the need to accommodate poor and marginalized people's need for subsistence products from forests, such as firewood and fodder. It turns out that the spillover effects between the two initiatives with synergistic outcomes need to be strengthened. Here we take them as an example of a *Goal–Goal* spillover effect: the REDD+ program aims at enhancing forest carbon stocks while the concurrent CFM focuses on multiple ecological and livelihood gains. However, those two programs have trade-offs. For example, CFM may stimulate local people to harvest trees to satisfy subsistence needs (Behavior 1), leading to decreases in forest carbon stocks (Goal 2), suggesting a *Behavior–Goal* negative spillover effect. Therefore, such trade-offs and the relevant mechanism must be carefully addressed to generate a win–win scenario.

8.10 Evidence from policy-mix

Empirical studies of green policy-mixes classify various policy interactions into several basic categories that include: complementary or synergistic, complementary when sequential, and redundant or conflicting (Gunningham & Sinclair, 1999; Robalino et al., 2015; Santos et al., 2015). Gunningham and Sinclair further describe a myriad of theoretical mixes that can occur between the following instruments: Command and control regulation, economic instruments, self-regulations, voluntarism, and information strategies (Gunningham & Sinclair, 1999). Below are some of the specific policy-mix examples that were described in the studies by Gunningham and Sinclair (1999), Barton et al. (2012), Robalino et al. (2015), and Santos et al. (2015). We refer to the Appendix for more information.

8.10.1 Complementary or synergistic

The 1990s witnessed the development and implementation of the US Environmental Protection Agency's 33/50 program, which encouraged relevant companies or organizations to reduce toxic chemical releases voluntarily. At the same time, existing command and control regulations for toxic chemical release remained in force. Therefore, the 33/50 program complemented the regulation policy by promoting more considerable reductions of toxic chemical release than the baseline, while companies were still required to comply with baseline levels. Similarly, the US Environmental Leadership Program provides regulatory relief for participating firms that go beyond compliance levels. Also, in the European Union (EU) there is consideration of compliance and inspection exemptions for firms that participate in eco-management and eco-audit schemes. In the US and EU examples, there is a backdrop of regulation that non-participating firms must follow. There is evidence regarding the effectiveness of the command and control regulation being complemented by voluntarism instruments (Gunningham & Sinclair, 1999).

In Australia, all vehicles built after 1985 were mandated to be fitted with catalytic converters, requiring the use of engines that ran on unleaded fuel.

Concurrently, the federal government introduced a phased price differential on the fuel price. In this context, leaded fuel became more expensive than unleaded fuel, which was an economic policy in the form of a pollution tax. While they are different approaches, these two policies complement one another because they provide the market with mutually supportive signals. The technology-based approach of requiring catalytic converters is directed at the manufacturer, and the pollution tax is aimed at the consumer (Gunningham & Sinclair, 1999).

8.10.2 Sequential relationship

Policy "sequencing" refers to certain instruments or policies being held in reserve and applied when another instrument fails or has serious problems (Gunningham & Sinclair, 1999). Here we present an example in Norway: the prime "command and control" instrument was the establishment of protected areas, primarily based on the appropriation of private land in biologically rich areas. The Trillemarka Nature Reserve, totaling 147 km^2 in size, was established as a conservation area of this kind with a relatively large size. Yet this instrument encountered opposition and conflict; it was superseded by a voluntary scheme with compensation payments. The command and control way of establishing protected areas is now almost dormant, and its future is quite uncertain, likely depending on the progress and results of the voluntary scheme (Barton et al., 2012).

In 2009, Norway passed the Nature Diversity Act, which includes—and integrates—all previous laws related to land use and biodiversity in one act. As the most important legal framework, this act stands as the fundamental guidelines for future regulatory and economic instruments in the domain of forest and biodiversity conservation—both inside and outside protected areas. The act provides guidelines for the management of priority species and selected habitat types and paves the way for Norway to fulfill its international commitments under the Convention on Biological Diversity, to which Norway is a signatory (Barton et al., 2012). An umbrella law or policy like this act will be instrumental in coordinating a variety of policies, foreseeing conflicts and optimizing synergism.

In many instances, an entire self-regulatory regime may not work well. Sequential interactions may come to help when economic incentives are imposed. In New Zealand, an industry-regulated program to reduce greenhouse gas emissions was introduced along with the announcement that, if the self-regulation failed, a carbon tax would be implemented. In Australia, a voluntary phase out of hydrochlorofluorocarbons (HCFCs) was legislated along with a call for a tradable quota policy: this policy would be implemented if the self-regulation failed (Gunningham & Sinclair, 1999).

8.10.3 Redundancies or conflicting policy interactions

Policy interactions may lead to a suboptimal—even negative—outcome when a command and control instrument is superimposed on an economic instrument, and both instruments target the same behavior (Gunningham & Sinclair, 1999).

In Costa Rica, a study was conducted to evaluate the effectiveness of two prevalent forest conservation policies that interact with each other: one is the policy of national parks and the other payments for ecosystem services (PES) programs; both focus on forest protection. The study area consists of park areas, park buffers with and without PES programs, and areas with PES programs far from parks. The study found a redundancy effect: the associated benefits of implementing parks and PES payments separately are greater than implementing them together. More specifically, it became more effective if one location was protected by a park and another by a payment than if one location was protected by both (Robalino et al., 2015).

Self-regulation and broad-based economic instruments may become incompatible, and here we show an example regarding the policy-mix used to phase out chlorofluorocarbon (CFCs) in Australia. As part of the National Ozone Strategy, the federal government imposed a cap on the production and importation of CFCs. Firms were allowed to trade CFC quotas under the condition that total CFCs were below the cap. After the inception of this program, federal and state governments brokered self-regulatory agreements with sector-specific industries to phase out the use of CFCs. This self-regulatory policy contradicted the cap, ultimately leading to the economic policy's failure (Gunningham & Sinclair, 1999).

In the USA, the Environmental Protection Agency's XL (eXcellence in Leadership) initiative was designed to give firms the flexibility of adopting less prescriptive, performance-oriented regulations. This policy was not successful partially because firms were concerned that the best available technology (BAT) regulations might still apply. Therefore, even if firms participated in an XL project, they might still be subject to penalties for failing to comply with the Clean Air and Clean Water Acts (Gunningham & Sinclair, 1999).

8.10.4 Economic instruments

Habitat Banking and Tradable Development Rights (TDR) stand as two beneficial economic instruments that play a role in the policy-mix sequential relationships. Such relationships include complementarities, redundancy, and conflicts with other instruments (Santos et al., 2015). Habitat banking aims to restore, create, enhance, or preserve off-site areas to provide compensatory mitigation for authorized impacts on habitats or biodiversity. A public agency, private organization, or landowner, rather than the developer, can establish conservation areas as mitigation for permitted impacts on biodiversity and ecosystems. The Wetland Mitigation Banking (the USA), Conservation Banking (the USA), New South Wales BioBanking (Australia), and BushBroker (Australia) are good examples of habitat banking schemes (Santos et al., 2015).

TDRs belong to a market-based approach, which aims to enhance land-use zoning by limiting land development and promoting biodiversity conservation. In designated areas, landowners are assigned TDRs as compensation for restricted development options, whereas in predicted growth areas, developers can choose to build at a baseline density or buy TDRs in order to realize a denser level of development. However, these methods may still generate ecological losses in return for the recreation or restoration of equivalent habitats. Both habitat banking and TDRs work in conjunction with a strong regulatory framework because the framework is necessary to ensure the adoption of the mitigation hierarchy and to determine the impacts to be offset. In most instances, habitat banking and TDR build on—and temporally follow—existing regulatory approaches to biodiversity offsetting and land-use zoning. For this reason, both instruments are characterized by sequential interactions or path dependence (Santos et al., 2015).

Under the Australian Threatened Species Legislation Amendment Act 2004, the New South Wales (NSW) BioBanking scheme was designed to support the biodiversity certification process. This scheme was consistent with the property vegetation planning process. It leverages provisions from other acts to ensure that the scheme and management actions are enforceable. Developers can choose between adopting the habitat banking scheme or negotiating an offset with the NSW government. The latter was their only option before introducing the BioBanking scheme. Developers are thus free to choose between offsetting the impact by themselves and purchasing the required credits. This kind of overlap promotes the flexibility and cost effectiveness of the overall policy-mix, likely leading to better achievement of conservation goals.

Habitat banking complements several other European Union policies, such as the Common Agricultural Policy and the Habitats Directive. Habitat banking may be instrumental in tackling the cumulative fragmentation of Europe's habitats by helping to restore, enlarge, and reconnect high nature value habitats. Yet the implementation of habitat banking and TDRs may give rise to the problem related to lack of additionality. If biodiversity outcomes would have occurred automatically as a result of existing instruments, such as management obligations set up for Natura 2000 sites, then people may ask why additional efforts were made for habitat banking and TDRs (Santos et al., 2015).

8.11 Summary of concurrent green initiatives

We present the descriptive data of all 15 cases, showing the country or continent, population size, area, urban or rural area, developed or developing countries or regions, funder type, and name of concurrent programs (Table 8.2). This summary table shows that concurrent green initiatives can be observed in most parts of the world regardless of the above variables. We also include empirical studies of green policy-mixes, which also point to the widespread existence of concurrent green initiatives and spillover effects among them.

Furthermore, we present the data sources regarding the population size and area of all the 15 cases in Tables 8.3 and 8.4, respectively. In many types of spillover effects, we have found both positive and negative effects—e.g., for *Behavior–Goal* spillover effect, we found a positive one in Australia and a negative one in Páramo (Section 8.2). In other instances, two elements can be achieved simultaneously (see the *Goal–Goal* spillover effects; Section 8.6) or must occur in sequence (Policy 1 and Policy 2 occur in sequence; Section 8.4). These findings

Table 8.2 Characteristics of the 15 selected cases[a]

Case name	Case ID	Country/continent	Population (area)	Urban–rural	Development level	Funder type[b]	Concurrent payment names[c]
USA	1	USA/N. America	332,639,102 (9,147,593 km²)	Urban and rural	Developed	Gov	EQIP and CRP
Jordan Lake	2	USA/N. America	719,888 (4,367 km²)	Urban and rural	Developed	Non-gov	Nitrogen credits vs. phosphorus credits
Neuse	3	USA/N. America	1,687,462 (15,700 km²)	Urban and rural	Developed	Gov	Wetland credits vs. nutrient credits
Yucatán and Chiapas	4	Mexico/N. America	7,315,083 (112,835 km²)	Rural	Developing to developed	Gov	PSA-H vs. PROCOREF
Páramo	5	Ecuador/S. America	N/A (18 km²)	Rural	Developing	Gov	PROFAFOR vs. SocioPáramo program
Rio Grande catchment	6	Bolivia/S. America	N/A (7,339 km²)	Rural	Developing	Gov	Level 1 payments vs. Levels 2 and 3 payments
New world[d] & Great Britain	7	Americas–Europe	63,200,000 (219,949 km²)	Urban and rural	Developing to developed	Gov	Carbon only vs. carbon-biodiversity payments
Baltic Sea	8	Denmark etc./Europe	85,000,000 (1,720,270 km²)	Urban and rural	Developed	Gov	Nitrogen payment vs. phosphorus payment
Marecchia & Foglia	9	Italy/Europe	404,800 (1,310 km²)	Urban and rural	Developed	Un-specified	Water retention vs. drinking water supply
Fanjingshan	10	China/Asia	21,000 (419 km²)	Rural	Developing	Gov	GTGP vs. FEBC
Tianma	11	China/Asia	17,295 (289 km²)	Rural	Developing	Gov	GTGP vs. FEBC
Wolong	12	China/Asia	5,950 (2,000 km²)	Rural	Developing	Gov	GTGB vs. NFCP

PVPF-KPWS	13	Cambodia/Asia	249,304+(5,925 km^2)[e]	Rural	Developing	Gov, Non-gov	Bird-nest program vs. ecotourism and agri-environment programs
Australia	14	Australia/Oceania	23,401,892 (7,741,220 km^2)	Urban and rural	Developed	Gov	Carbon only vs. carbon-biodiversity payments
Nepal	15	Nepal/Asia	26,490,000 (147,181 km^2)	Rural	Developing	Gov, Non-gov	REDD+ vs. CFM

Notes:

[a] For sources of information, see Tables 8.3 and 8.4.

[b] "Gov" and "Non-gov" refer to governmental and non-governmental sponsors, respectively.

[c] For information about acronyms, see the following text.

[d] New World refers to the Americas and Oceania; the numbers 63,200,000 (219,949 km^2) are only about Great Britain (the UK) based on the source.

[e] The data for Population in Kulen Promtep (KP) Wildlife Sanctuary is unknown; the numbers are only for Preah Vihear (PV).

Table 8.3 Sources of population information in Table 8.2

Site name	Population reference
Fanjingshan	Wandersee, S.M., An, L., López-Carr, D., & Yang, Y. (2012). Perception and decisions in modeling coupled human and natural systems: A case study from Fanjingshan National Nature Reserve, China. Ecological Modelling, 229, 37–49.
Wolong	Xu, J., Wei, J., & Liu, W. (2019). Escalating human–wildlife conflict in the Wolong Nature Reserve, China: A dynamic and paradoxical process. Ecology and Evolution, 9(12), 7273–7283.
Tianma	Zhang, Q., Bilsborrow, R. E., Song, C., Tao, S., & Huang, Q. (2019). Rural household income distribution and inequality in China: Effects of payments for ecosystem services policies and other factors. Ecological Economics, 160, 114–127.
Neuse	State of North Carolina, North Carolina Department of Environment and Natural Resources, & Office of Environmental Education and Public Affairs. (2013). *Neuse River basin.* NC: North Carolina Department of Environment and Natural Resources, & Office of Environmental Education and Public Affairs. Retrieved from https://files.nc.gov/deqee /documents/files/neuse.pdf
Jordan Lake	NC Geographic Information Coordinating Council (GICC). (2018). *2000_Census_Blocks* [Shapefile]. Retrieved from https://hub.arcgis.com/datasets/nconemap::2000-census-blocks
Páramo	n/a
Marecchia & Foglia	Morri, E., Pruscini, F., Scolozzi, R., & Santolini, R. (2014). A forest ecosystem services evaluation at the river basin scale: Supply and demand between coastal areas and upstream lands (Italy). Ecological indicators, 37, 210–219.
Baltic Sea	Lääne, A., Kraav, E., Titova, G., United Nations Environment Programme (UNEP). (2005). *Global International Waters Assessment: Baltic Sea, GIWA Regional assessment 17.* Kalmar, Sweden: University of Kalmar.
Australia	Australian Bureau of Statistics. (2019). *2016 Census QuickStats.* Retrieved from https://quickstats.censusdata.abs.gov.au /census_services/getproduct/census/2016/quickstat/036
New World & Great Britain	United Nations Department of Economic and Social Affairs. (2018). *2017 Demographic Yearbook.* NY: United Nations. Office for National Statistics. (2012). *2011 Census: Population Estimates for the United Kingdom, March 2011.* Retrieved from https://www.ons.gov.uk/peoplepopulationandcommunity/populationandmigration/populationestimates /bulletins/2011censuspopulationestimatesfortheunitedkingdom/2012-12-17
Preah Vihear & Kulen Promtep	OCHA ROAP. (2018). khm_pop_2016_adm3_v2 [csv]. Retrieved from https://data.humdata.org/dataset/cambodia -population-statistics

Rio Grande Valles — Instituto Nacional de Estadistica. (2012). Resultados: Censo de Poblacion y Vivienda 2012. Retrieved from http://datos.ine.gob.bo/binbol/RpWebEngine.exe/Portal?BASE=CPV2012COM&lang=ESP

USA — Umited States Census Bureau. (2019). Population estimates, July 1, 2019, (V2019) – United States [data table]. QuickFacts. Retrieved from https://www.census.gov/quickfacts/fact/table/US

Yucatán and Chiapas — Instituto Nacional de Estadistica, Geografia e Informática (INEGI). (2015). Encuesta Intercensal 2015: Principales Resultados. Retrieved from https://www.inegi.org.mx/contenidos/programas/intercensal/2015/doc/eic_2015_presentacion.pdf

Nepal — CBS, 2011. National Population and Housing Census 2011 (National Report). Central Bureau of Statistics (Nepal), June 22, 2011. https://web.archive.org/web/20130418041642/http://cbs.gov.np/wp-content/uploads/2012/11/National%20Report.pdf

Table 8.4 Sources of area information in Table 8.2

Site name	Area reference
Fanjingshan	Tsai, Y.H., Stow, D., Chen, H.L., Lewison, R., An, L., & Shi, L. (2018). Mapping vegetation and land-use types in Fanjingshan National Nature Reserve using Google Earth Engine. Remote Sensing, 10(6), 927.
Wolong	An, L., He, G., Liang, Z., & Liu, J. (2006). Impacts of demographic and socioeconomic factors on spatio-temporal dynamics of panda habitat. Biodiversity & Conservation, 15(8), 2343–2363.
Tianma	Xu, J.L., Zhang, Z.W., Zheng, G.M., Zhang, X.H., Sun, Q.H., & McGowan, P. (2007). Home range and habitat use of Reeves's Pheasant Syrmaticus reevesii in the protected areas created from forest farms in the Dabie Mountains, central China. Bird Conservation International, 17(4), 319–330.
Neuse	State of North Carolina, North Carolina Department of Environment and Natural Resources, & Office of Environmental Education and Public Affairs. (2013). *Neuse River basin*. NC: North Carolina Department of Environment and Natural Resources, & Office of Environmental Education and Public Affairs. Retrieved from https://files.nc.gov/deqee/documents/files/neuse.pdf
Páramo	We only used data of Mazar Wildlife Reserve within Páramo. *Mazar Wildlife Reserve (MWR) general information*. Retrieved from https://docplayer.net/217990-Mazar-wildlife-reserve-mwr-general-information.html
Marecchia & Foglia	Morri, E., Pruscini, F., Scolozzi, R., & Santolini, R. (2014). A forest ecosystem services evaluation at the river basin scale: Supply and demand between coastal areas and upstream lands (Italy). Ecological indicators, 37, 210–219.
Baltic Sea	Lääne, A., Kraav, E., Titova, G., United Nations Environment Programme (UNEP). (2005). *Global International Waters Assessment: Baltic Sea, GIWA Regional assessment 17*. Kalmar, Sweden: University of Kalmar.
Australia	United Nations Department of Economic and Social Affairs. (2018). *2017 Demographic Yearbook*. NY: United Nations.
New World	United Nations Department of Economic and Social Affairs. (2018). *2017 Demographic Yearbook*. NY: United Nations.
Great Britain	United Nations Environment Programme (UNEP). (1998). *Island directory: Islands of United Kingdom*. Retrieved from http://islands.unep.ch/ICJ.htm
Preah Vihear & Kulen Promtep	Clements, T., John, A., Nielsen, K., An, D., Tan, S., Milner-Gulland, E.J. (2010). Payments for biodiversity conservation in the context of weak institutions: Comparison of three programs from Cambodia. Ecological Economics, 69(6), 1283–1291.
Rio Grande catchment	Pynegar, E.L., Jones, J.P., Gibbons, J.M., & Asquith, N.M. (2018). The effectiveness of Payments for Ecosystem Services at delivering improvements in water quality: lessons for experiments at the landscape scale. PeerJ, 6, e5753.

USA	Umited States Census Bureau. (2018). State Area Measurements and Internal Point Coordinates [data table]. Geographies, Reference Files. Retrieved from https://www.census.gov/geographies/reference-files/2010/geo/state -area.html
Yucatán and Chiapas	Instituto Nacional de Estadística, Geografía e Informática (INEGI). (2013). Información por entidad: Yucatán. Retrieved from https://web.archive.org/web/20130531052605/http://cuentame.inegi.gob.mx/monografias/informacion/yuc/ default.aspx?tema=me&e=31 (for Yucatán) Instituto Nacional de Estadística, Geografía e Informática (INEGI). (2013). Información por entidad: Chiapas. Retrieved from https://web.archive.org/web/20130602020035/http://cuentame.inegi.gob.mx/monografias/informacion/chis/ default.aspx?tema=me&e=07 (for Chiapas)
Nepal	Anup, K. C., Rijal, K., & Sapkota, R. P. (2015). Role of ecotourism in environmental conservation and socioeconomic development in Annapurna conservation area, Nepal. International Journal of Sustainable Development & World Ecology, 22(3), 251–258. https://doi.org/10.1080/13504509.2015.1005721

are exciting, yet no systematic research has been devoted to such spillover effects and the underlying mechanisms.

Appendix: Policy-mix examples

This appendix contains seven cases where two policies or more have spillover effects among one another. Given the text length and detail, we list them as cases rather than put them in a table. In each case, we identify concurrent PES policies (Policy A, Policy B, and Policy C if any), point out the spillover type, present some detail, and give the reference(s).

Case 1: Australia policy-mix

Policy "A"—The New South Wales (NSW) BioBanking (2008) is a habitat banking scheme that uses provisions from other acts to ensure that scheme and management actions are enforceable and consistent with the property vegetation planning process.

Policy "B"—Australian Threatened Species Legislation Amendment Act 2004 incorporates a biodiversity certification process in order to protect threatened species.

Spillover type: Policy "A" *supports* and *complements* Policy "B".

Details: The banking scheme was designed for that purpose. Developers can choose between adopting the habitat banking scheme or negotiating an offset with the NSW government. The latter was their sole option prior to introducing the BioBanking scheme. Developers are thus free to choose between offsetting the impact themselves and purchasing the required credits. This kind of overlap heightens the flexibility and cost effectiveness of the overall policy-mix to achieve conservation goals.

Source: Santos, R., Schröter-Schlaack, C., Antunes, P., Ring, I. & PEDRO Clemente, P. (2015). Reviewing the role of habitat banking and tradable development rights in the conservation policy-mix. Environmental Conservation. FirstView Article 1-12.

DOI: 10.1017/S0376892915000089 Published online: Apr 2015.

Case 2: The USA policy-mix

Policy "A"—US EPA's Project XL—1995 initiative designed to promote cleaner technologies by giving firms flexibility by adopting less prescriptive, performance-oriented regulations.

Policies "B"—The Clean Air Act—1963 federal law designed to control air pollution on a national level.

Policy "C"—The Clean Water Act—1977 primary federal law in the USA governing water pollution.

Spillover type: Project XL is a major *contradiction* to the Clean Air and Clean Water Acts.

Details: Even if firms participated in Project XL, which calls for more relaxed standards, they might still be prosecuted for failing to comply with the best available technology standards established by the Clean Air and Clean Water Acts. Ultimately, Project XL failed.

Source: Gunningham, N. & Sinclair, D. (1999). Regulatory Pluralism: Designing Policy Mixes for Environmental Protection. *Law & Policy*, Vol, 21. No.1. 49–75.

Case 3: The USA policy-mix

Policy "A"—historical regulatory instruments that established protected areas such as Trillemarka Nature Reserve in Norway created on December 13, 2002 (date found on Wikipedia). This type of command and control regulation was met with opposition and conflict.

Policy "B"—Voluntary Conservation Approach—an economic instrument was proposed in 2000. Since 2003, nearly all new conservation processes have been in this voluntary form. Forest owners with biodiversity hotspots on their property can receive compensation for protecting areas as a nature reserve.

Policy "C"—Norway Nature Diversity Act established in 2009—oversees all previous and current laws related to land use and biodiversity and is a legal framework for all future *regulatory* and *economic instruments* in forest and biodiversity.

Spillover type: The voluntary conservation approach ("B") has a *sequential complementary relationship* with the historical regulatory instrument ("A").

Details: Policy "sequencing" occurs when certain instruments are applied when another instrument fails or has shortcomings. The Nature Diversity Act ("C") *complements* the previous policies because it *coordinates* these and other policies.

Source: Barton, D.N., Lindhjem, H., Rusch, G.M., Sverdrup-Thygeson, A., Blumentrath, S., Sørheim, M.D., Svarstad, H., & Gundersen, V. (2012). *Assessment of Existing and Proposed Policy Instruments for Biodiversity Conservation in Norway*. POLICYMIX Report Issue No 1/2012. Oslo, Norway.

Case 4: Nepal policy-mix

Policy "A"—In Nepal, Chitwan National Park (CNP) was established in 1973 as a protected, regulated area managed by the Department of National Parks and Wildlife Conservation. Aside from a 3-day grass collecting period, resource collection is prohibited in the park.

Policy "B"—Incentive-based programs (IBPs). These programs can empower and provide skill training for local people, while developing revenue sharing programs, sustainable extraction programs, and tourist markets.

Spillover type: IBPs (Policy "B") was set up to *complement* the CNP. However, villagers' actions do not always coincide with the views that they express about the importance of conservation.

Details: Residents surrounding CNP continue to disregard legal restrictions on resource collection. One goal is to create a link or association between the social/economic benefits and conservation efforts. Through surveys, it was determined that there had been some successes as a result due to the combination of these programs and policies. However, there is an inability to deliver benefits to the population surrounding the park. Villagers far from the entry point get fewer benefits than the gateway village.

Source: Nepal, S., & Spiteri, A. (2011). Linking Livelihoods and Conservation: An Examination of Local Residents' Perceived Linkages Between Conservation and Livelihood Benefits Around Nepal's Chitwan National Park. *Environmental Management*. 47:727–738.

Case 5: European Union policy-mix

Policy "A"—The European Union's Common Agricultural Policy (specifical provisions under more recent CAP reforms). This policy provides essential economic support to farmers sustainably managing wood pastures through direct payments to low-intensity livestock farmers for the variety of ecosystem services they provide.

Policy "B"—The EU's Rural Development Policy provides payments to wood-pasture farmers and others who go above and beyond environmental standards established under the CAP.

Policy "C"—Natura 2000—at the core of EU Habitats Directive—maintains and restores natural habitats.

Spillover type: The Rural Development Policy *supplements* the CAP provisions, yet *contradicts* the CAP because it establishes agro-forestry systems on agricultural land, some of which could be woody pastures. The Natura 2000 and EU Habitat Directive seem to directly *contradict* the CAP.

Details: Of the 233 natural habitat types included in this directive, 65 have some relationship to wood pastures, yet many are referred to as forest habitats. The criteria for forest habitats under Natura 2000 call for the restoration of tall, ungrazed, dense forests which do not allow sustainable livestock grazing in forests and do not safeguard wood pastures.

Source: Plieninger, T., Hartel, T., Martín-López, B., Beaufoy, G., Bergmeier, E., Kirby, K., Jesús Monterog, M., Moreno, G., Oteros-Rozas, E., and Van Uytvanck, J. (2015). Wood-pastures of Europe: Geographic coverage, social–ecological values, conservation management, and policy implications. *Biological Conservation*. 190: 201570–79.

Case 6: Brazilian policy-mix

Policy "A"—The Brazilian Forest Code, a federal law, establishes a percentage of the area of rural properties that are to be maintained as a permanent forest reserve. As of 1996, deforestation was prohibited on 80% of private land-holdings in the "Legal Amazon" region.

Policy "B"—Ecological-Economic Zone (EEZ)—provides allocation of credit and other public incentives and allows the reserved area to be reduced to 50% in designated, productive-use areas that are involved in the EEZ. Forests may be managed for timber and non-timber production/extraction.

Spillover type: While the Brazilian Forest Code maintains a baseline for conservation requirements, the EEZ *supplements* the Code by creating allocations of credit and public incentives in productive-use areas.

Details: Other policies that complement these efforts are the 1998 Environmental Crime Law that enforces conservation efforts and streamlines court proceedings and Integrated System for Monitoring and Licensing (SIMLAM), an environmental monitoring system that integrates satellite images and forest inspections. A rural credit system is offered to landowners who register with SIMLAM, and some government banks require a declaration of compliance with the Forest Code to be screened for credit.

Source: May, P.H., Andrade, J., Vivan, J.L., Kaechele, K, Fernanda Gebara, M., and Abad, R. (2012). *Assessment of the role of economic and regulatory instruments in the conservation policymix for the Brazilian Amazon – a coarse grain analysis.* POLICYMIX Report. Issue No 5/2012. Oslo, Norway.

Case 7: Indonesia policy-mix

Policy "A"—Reducing Emissions from Deforestation and Forest Degradation (REDD+)—adopted in 2010 by the United Nations Framework Convention on Climate Change (UNFCC)—conserves forests, enhances forest carbon stocks, and sustainably manages forests.
Policy "B"—Kyoto Protocol—agroforest projects.

Spillover type: These two policies did not support one another. There were trade-offs between carbon sequestration and biodiversity.

Details: The above results were based on a carbon and biodiversity management study in Sulawesi, Indonesia. Contradictions between REDD+ projects and Kyoto Protocol agroforest projects need to be explored further.

Source: Kessler, M., Hertel, D., Jungkunst, H., Kluge, J., Abrahamczyk, S., et al. (2012). Can joint carbon and biodiversity management in tropical agroforestry landscapes be optimized? *PLOS ONE*, 7(10), e47192–e47196.

References

Barton, D. N., Lindhjem, H., Sverdrup-Thygeson, A., Blumentrath, S., Sørheim, M. D., Svarstad, H., & Gundersen, V. (2012). *Assessment of existing and proposed policy instruments for biodiversity conservation in Norway.* NINA. http://policymix.nina.no/

Bastakoti, R. R., & Davidsen, C. (2014). REDD+ and forest tenure security: Concerns in Nepal's community forestry. *International Journal of Sustainable Development & World Ecology, 21*(2), 168–180. https://doi.org/10.1080/13504509.2013.879542

Bottazzi, P., Wiik, E., Crespo, D., & Jones, J. P. G. (2018). Payment for environmental "self-service": exploring the links between farmers' motivation and additionality in a conservation incentive programme in the Bolivian Andes. *Ecological Economics*, *150*, 11–23. https://doi.org/10.1016/j.ecolecon.2018.03.032

Bremer, L. L., Farley, K. A., Chadwick, O. A., & Harden, C. P. (2016). Changes in carbon storage with land management promoted by payment for ecosystem services. *Environmental Conservation*, *43*(4), 397–406. Cambridge Core. https://doi.org/10.1017/S0376892916000199

Bryan, B. A., Runting, R. K., Capon, T., Perring, M. P., Cunningham, S. C., Kragt, M. E., Nolan, M., Law, E. A., Renwick, A. R., Eber, S., Christian, R., & Wilson, K. A. (2015). Designer policy for carbon and biodiversity co-benefits under global change. *Nature Climate Change*, *6*, 301.

Clements, T., John, A., Nielsen, K., An, D., Tan, S., & Milner-Gulland, E. J. (2010). Payments for biodiversity conservation in the context of weak institutions: Comparison of three programs from Cambodia. *Special Section: Payments for Environmental Services: Reconciling Theory and Practice*, *69*(6), 1283–1291. https://doi.org/10.1016/j.ecolecon.2009.11.010

Devkota, B. P. (2020). Social inclusion and deliberation in response to REDD+ in Nepal's community forestry. *Forest Policy and Economics*, *111*, 102048.

Devkota, B. P., & Mustalahti, I. (2018). Complexities in accessing REDD benefits in community forestry: Evidence from Nepal's Terai region. *International Forestry Review*, *20*(3), 332–345. https://doi.org/10.1505/146554818824063041

Eckholm, E. P. (1975). The deterioration of mountain environments. *Science*, *189*(4205), 764–770.

Ezzine-de-Blas, D., Dutilly, C., Lara-Pulido, J.-A., Velly, G. L., & Guevara-Sanginés, A. (2016). Payments for environmental services in a policymix: Spatial and temporal articulation in Mexico. *PLoS ONE*, *11*(4), e0152514.

Gilmour, D. (2016). *Forty years of community-based forest management*. FAO, FAO Forestry Paper (FAO) Eng No. 176.

Gren, I.-M., & Elofsson, K. (2017). Credit stacking in nutrient trading markets for the Baltic Sea. *Marine Policy*, *79*, 1–7. https://doi.org/10.1016/j.marpol.2017.01.026

Gunningham, N., & Sinclair, D. (1999). Regulatory pluralism: Designing policy mixes for environmental protection. *Law & Policy*, *21*(1), 49–76. https://doi.org/10.1111/1467-9930.00065

Maraseni, T., Neupane, P. R., Lopez-Casero, F., & Cadman, T. (2014). An Assessment of the impacts of the REDD+ pilot project on community forest user groups (CFUGs) and their community forests in Nepal. *Journal of Environmental Management, 136*, 37–46.

Marquardt, K., Khatri, D., & Pain, A. (2016). REDD+, forest transition, agrarian change and ecosystem services in the hills of Nepal. *Human Ecology*, *44*, 229–244. https://doi.org/10.1007/s10745-016-9817-x

Morri, E., Pruscini, F., Scolozzi, R., & Santolini, R. (2014). A forest ecosystem services evaluation at the river basin scale: Supply and demand between coastal areas and upstream lands (Italy). *Ecological Indicators*, *37*, 210–219. https://doi.org/10.1016/j.ecolind.2013.08.016

Motallebi, M., Hoag, D. L., Tasdighi, A., Arabi, M., Osmond, D. L., & Boone, R. B. (2018). The impact of relative individual ecosystem demand on stacking ecosystem credit markets. *Ecosystem Services*, *29*, 137–144.

Newton, P., Schaap, B., Fournier, M., Cornwall, M., Rosenbach, D. W., DeBoer, J., Whittemore, J., Stock, R., Yoders, M., Brodnig, G., & Agrawal, A. (2015). Community forest management and REDD+. *Forest Policy and Economics*, *56*, 27–37. https://doi .org/10.1016/j.forpol.2015.03.008

Phelps, J., Webb, E. L., & Agrawal, A. (2010). Does REDD+ threaten to recentralize forest management? *Science*, *328*(5976), 312–313. https://doi.org/10.1126/science .1187774

Poudel, M., Thwaites, R., Race, D., & Dahal, G. R. (2014). REDD+ and community forestry: Implications for local communities and forest management: A case study from Nepal. *International Forestry Review*, *16*(1), 39–54. https://doi.org/10.1505 /146554814811031251

Program Evaluation Division. (2009). *Department of Environment and Natural Resources Mitigation Credit Determinations* [Special Report to the General Assembly Report Number 2009-04]. North Carolina General Assembly. https://www.ncleg.net/PED/ Reports/documents/Wetlands/Wetland_Report.pdf

Robalino, J., Sandoval, C., Barton, D. N., Chacon, A., & Pfaff, A. (2015). Evaluating interactions of forest conservation policies on avoided deforestation. *PLOS ONE*, *0124910*, 1–16. https://doi.org/.

Robertson, M., BenDor, T. K., Lave, R., Riggsbee, A., Ruhl, J., & Doyle, M. (2014). Stacking ecosystem services. *Frontiers in Ecology and the Environment*, *12*(3), 186–193. https://doi.org/10.1890/110292

Santos, R., Schröter-Schlaack, C., Antunes, P., Ring, I., & Clemente, P. (2015). Reviewing the role of habitat banking and tradable development rights in the conservation policy mix. *Environmental Conservation*, *42*(4), 294–305. https://doi.org/10.1017/ S0376892915000089

Sharma, B. P., Karky, B. S., Nepal, M., Pattanayak, S. K., Sills, E. O., & Shyamsundar, P. (2020). Making incremental progress: Impacts of a REDD+ pilot initiative in Nepal. *Environmental Research Letters*, *15*(10), 105004. https://doi.org/10.1088/1748-9326/ aba924

Shrestha, S., Shrestha, U. B., & Bawa, K. S. (2017). Contribution of REDD+ payments to the economy of rural households in Nepal. *Applied Geography*, *88*, 151–160. https://doi .org/10.1016/j.apgeog.2017.09.001

Song, C., Bilsborrow, R., Jagger, P., Zhang, Q., Chen, X., & Huang, Q. (2018). Rural household energy use and its determinants in China: How important are influences of payment for ecosystem services vs. Other factors? *Ecological Economics*, *145*, 148–159. https://doi.org/10.1016/j.ecolecon.2017.08.028

Thomas, C. D., Anderson, B. J., Moilanen, A., Eigenbrod, F., Heinemeyer, A., Quaife, T., Roy, D. B., Gillings, S., Armsworth, P. R., & Gaston, K. J. (2013). Reconciling biodiversity and carbon conservation. *Ecology Letters*, *16*(s1) Supplement 1, 39–47. https://doi.org/10.1111/ele.12054

Yang, W., Lupi, F., Dietz, T., & Liu, J. (Jack). (2016). Dynamics of economic transformation. In J. Liu, V. Hull, W. Yang, A. Viña, X. Chen, Z. Ouyang, H. Zhang (Eds.), *Pandas and people: Coupling human and natural systems for sustainability* (pp. 109–119). Oxford University Press.

Yost, A., An, L., Bilsborrow, R., Shi, L., Chen, X., & Zhang, W. (2020). Linking concurrent payments for ecosystem services in a Chinese nature reserve. *Ecological Economics*, *169*, 106509.

9 Conclusions

Ecosystems are being degraded or destroyed at an alarming rate, jeopardizing their vital services, including food, water, clean air, soil, and forests (Daily & Matson, 2008; Millennium Ecosystem Assessment, 2005). In response, various green initiatives, including payments for environmental services, have been developed and implemented for decades, providing incentives to various stakeholders (e.g., ecosystem service providers, landowners, or users in most cases) to take actions or to refrain from harmful actions. In the domain of generic conservation policy, scholars have started to consider policy coordination among a mix of policies (i.e., a policy-mix), which may include command-and-control (e.g., protected areas), economic incentives in the form of cash transfers (e.g., payments for environmental services), and capacity building (e.g., forest management support and community enterprises) policies. Surprisingly, concurrent green initiatives, especially the spillover effects, have seldom been addressed. Consequently, green initiatives are rarely coordinated with one another despite the increasingly recognized need to do so. In this book, we collected data from sites or countries worldwide to demonstrate the urgency of this largely overlooked phenomenon.

9.1 Green initiatives worldwide

We examined empirical data for concurrent active green initiatives worldwide in this book. The geographic scopes of these green initiatives range from global scale (e.g., the Green Climate Fund and the REDD+ program) to continental scale (e.g., European Green Deal) to national scale (e.g., the CRP in the USA and the GTGP in China) and local scale (e.g., the Jordan Lake water quality in North Carolina, USA). Many of the green initiatives we identify have concurrent green initiative(s) that are being implemented simultaneously at the same location and/or benefiting the same people. Each of these green initiative policies aims to change some behaviors of the stakeholders to gain the desired outputs. The changes in the behaviors range from increase or maintenance of an existing positive behavior (e.g., planting more trees or maintenance of existing tree cover), to the reduction of harmful behavior (e.g., using fewer fertilizers or pesticides), to stimulating a new positive behavior (e.g., diversifying livelihood options to reduce reliance on

DOI: 10.4324/9781003290292-9

natural resources). Diverse policy environmental outcomes are sought from these green initiatives, including enhancement of carbon sequestration (e.g., the global REDD+ program) for global warming mitigation, habitat preservation for biodiversity conservation (e.g., Australian environmental planting program), soil and water conservation (e.g., China's GTGP program and the US CRP program), and water quality improvement (e.g., the US EQIP and the N and P reduction program in the Baltic Sea).

The *Policy–Behavior–Goal* scheme designed in one policy may have hidden spillover effects on the *Policy–Behavior–Goal* of another concurrent green initiative during policy development and implementation stages. Some hidden spillover effects are synergistic, i.e., implementation of one green initiative helps the other initiative achieve its goals, whereas other hidden spillover effects are deleterious, i.e., implementation of one green initiative compromises the other green effort achieving its goals. Moreover, such interactions are multi-faceted and complex. Both synergistic and deleterious interactions may exist among the same concurrent green initiatives. We must identify how these interactions happen to design new green initiatives that maximize the synergies among the existing green initiatives or cost-effectively fine-tune the existing concurrent green initiatives to gain the best desired environmental outcomes.

In the 26th Conference of Parties (COP 26) of the United Nations Framework Convention on Climate Change in Glasgow, Scotland, in 2021, the party recognized global warming as an existential threat to humanity. Many countries pledged more green initiatives to reduce carbon emissions or enhance carbon sequestration. The European Union rolled out the European Green Deal, which aims at reducing carbon emissions by 50% by 2030 and net-zero carbon emissions by 2050. The US government made a similar pledge, i.e., 50% carbon emission reduction by 2030 and net-zero carbon emission by 2050. A New Green Deal bill had been proposed in the US Congress to address climate change and sustainable development. Although the New Green Deal has not passed the US Congress as we write this book, the passage of such a bill in one name or another is a matter of time because there is no viable alternative sustainability mechanism. Unlike the previous 25 meetings, the developing countries also pledged to the carbon reduction target in COP 26. Two of the largest carbon emitters in the developing world, China and India, pledged to reach net-zero carbon emissions by 2060 and 2070, respectively. Moreover, more than 130 countries pledged to stop deforestation by 2030 at the 2021 global climate summit.

We can expect many new green initiatives in the coming decades to achieve the goals of the pledges made by various countries. These new green initiatives will inevitably interact with the existing green initiatives. The mechanisms of spillover effects among concurrent green initiatives we identify in this book are vital to designing cost-effective new green initiatives. By maximizing the synergistic spillover effects among the new green initiatives and minimizing deleterious ones, policymakers can design new green initiatives to achieve faster and cheaper emission reduction goals.

9.2   Losses and gains in concurrent green initiatives

The no-spillover presumption is prevalent in an era of global change when "[t]he biosphere, upon which humanity as a whole depends, is being altered to an unparalleled degree across all spatial scales", according to the most recent IPBES report (IPBES, 2019). One example of such global change is species extinction, which is occurring at an alarming rate that far outpaces losses in the fossil record, which—if not averted—will likely lead to the Earth's sixth mass extinction (Hooper et al., 2012). In response, governments and non-government organizations have committed considerable resources worldwide to combat such global change through various agreements such as the Convention on Biological Diversity (CBD) and United Nations Sustainable Development Goals (SDGs). Under the no-spillover effect presumption, conservation efforts have rarely been coordinated among programs and/or agencies despite the increasingly recognized need (Barton et al., 2013; Ezzine-de-Blas et al., 2016). This defect, and other deficiencies, such as inadequate finance and poor institutional arrangements, may account for many "non-optimistic outcomes" of the SDG and CBD endeavors (Quétier et al., 2014). For instance, the 2010 CBD target was not achieved (Waldron et al., 2017), and the 2020 Aichi biodiversity targets were not accomplished (Secretariat of the Convention on Biological Diversity, 2020, p. 5).

Our conceptual green initiative analytical framework (Figure 1.3) aims to uncover spillover effects among concurrent payments for ecosystem services, improving the effectiveness of conservation payments, programs, or policies. Our analysis under this framework testifies against the prevalent no-spillover effect presumption, suggesting that essential relationships exist among concurrent green initiatives—and very likely among many (if not all) other types of concurrent conservation efforts worldwide. With such hidden spillover effects made explicit, scientists, conservationists, policymakers, and other stakeholders will be able to track the cascading co-benefits or hidden losses of specific conservation policies or payments. For instance, implementing FEBC payments has likely generated 8.25% and 8.87% more GTGP land enrollment between 2000 and 2010 at Fanjingshan and Tianma reserves, respectively (Section 5.4). Extrapolating the average rate (8.56%) to the whole country due to their spatial concurrency in China (Table 4.1), China should have gained 6.93 million ha "additional" GTGP farmland as an FEBC-induced co-benefit, which could have translated (conservatively) to 1,435,850 million metric tons of carbon sequestration per year.

However, if implementing a hypothetical GTGP policy (see Section 8.8 for *Time–Time* spillover effects), the FEBC payments will likely reduce possible enrollment in GTGP by 0.1653 million ha, corresponding to a reduction in carbon sequestration of 503,173 million metric tons (Section 5.4). In the USA, the Environmental Quality Incentives Program (EQIP) and the Conservation Reserve Program (CRP) have coexisted since 2002 as concurrent payments for environmental services (PES) programs (Chapter 3), bearing a considerable overlap in their goals: preserving water, soil, and wildlife habitat. Between 2020 and 2022, a total of 12.4 million acres of CRP land will expire (United States Department of Agriculture & Farm Service Agency, 2019), accounting for a total of US$1.01 billion in annual CRP payments assuming that a total yearly CRP investment of US$2 billion (USDA Farm Service Agency, 2016) and CRP acreage of 22,609,492 (United States Department of Agriculture & Farm Service Agency, 2019). However, the US House Agriculture Committee passed the 2018 Farm Bill, which "allows a landowner to enroll in EQIP during the last year of a CRP contract". Aside from ignoring a reported offsetting spillover effect (i.e., EQIP payments decrease CRP participation), this decree might also open the potential for "double-dipping" as occurred in the Neuse (North Carolina, USA) case.

Spillover effects between green initiatives appear to be prevalent globally (Tables 1.1–1.3), generating substantial negative and positive consequences. Surprisingly, such effects were primarily ignored in green initiative policymaking and implementation. In both academia and conservation practice arenas, there have been many calls to explore the connections between policies (e.g., policy-mix and policyscape (Ezzine-de-Blas et al., 2016)), between different regions (e.g., telecoupling (Liu et al., 2013)), and between multiple goals (e.g., impacts of the intervention on non-targeted services (Naeem et al., 2015)). There existed literature to call for links between policy designs (e.g., avoiding oversimplified design and implementation (Wunder et al., 2018)) and the bundling and stacking of relevant payments for environmental services (Gren & Elofsson, 2017; Program Evaluation Division, 2009). Surprisingly, concurrent green initiatives are rarely coordinated with one another (Barton et al., 2013; Ezzine-de-Blas et al., 2016).

Examining and leveraging such spillover effects should uncover hidden losses or undocumented co-benefits. Doing so may help reduce the negative impacts on the environment due to budget cuts. If scientists can identify significant co-benefits of green initiatives such as the Green New Deal, lawmakers may have better standing to defend them. Similarly, green initiatives with significant negative spillover effects can be successfully suspended.

9.3 Why is there no attention to spillover effects

The no-spillover presumption, along with the corresponding green practice, has surprisingly coexisted for a long time with an emerging literature that calls for exploring connections between policies (e.g., policy-mix and policyscape (Ezzine-de-Blas et al., 2016)), between geographic areas (e.g., telecoupling (Liu et al., 2013)), and between PES designs (e.g., bundling and stacking (Gren & Elofsson, 2017); Program Evaluation Division, 2009).

The existing literature does not fully recognize the existence of concurrent green initiatives (e.g., PES programs), although policies (regarding payment schemes) are usually embedded with other policy tools known as policy-mixes (Börner et al., 2017; Yost et al., 2020). The spatial representation and expression of a certain policy-mix, named a policyscape, have received attention in the last decade because the capacity of a policy-mix to achieve various goals depends on the degree to which policies within the policy-mix align with one another spatially

(Barton et al., 2013). Barton et al. used a case in Norway to assess the spatial distribution of forest policies and argue why such a spatial representation can help more efficient planning. Ezzine-de-Blas et al. further developed the concept of *policyscape* to study the spatial and temporal articulation of the Mexican PES with other policies such as agricultural incentives and protected areas (Ezzine-de-Blas et al., 2016). They found some coordination between projects within the same agency but little coordination between agencies.

The paucity of studies on exploring and testing the spillover effects among concurrent green initiatives may stem from the following three aspects: the lack of general framework, the lack of policy design with more than one tool taken into account, and the lack of data and methodologies for examining such concurrent programs.

First, it is in dire need to review the existing knowledge on concurrently implemented programs to better frame and test the theoretical understanding of the concurrent programs with their spillover effects. Previous studies have rarely recognized the hidden spillover effects between the policies, the behavioral changes the policies intend to incentivize, and the potential achievement of goals or gains resulting from the changes in policies and/or behavioral patterns. The interactions across institutional, socioeconomic, and ecological scales are largely missing in the current literature for evaluating the policy outcomes, making the modeling effort incomplete.

Second, there is a lacuna in designing concurrent policies or programs from a governmental perspective. Policymakers often treated one single policy as a tool independent from others; even these programs targeted the same regions or involved the same group of recipients. For instance, in GTGP and FEBC, the design of the enrollment of farmland to forest under the GTGP seldom considered how farmers change their activities to use natural forests under the protection of FEBC in adaptation to the loss of farmland. Case studies in Tianma and Fanjingshan (China) found strong evidence that participation in FEBC increases the likelihood of participation in GTGP (Chapters 5 and 6). Furthermore, few studies adopted methodologies specific to the investigation of the interrelationships between concurrent programs, making available data scarce to examine spillover effects. The lack of panel datasets with baseline information for policy evaluation likely increases the difficulty of assessing more than one program. Although spatial data such as satellite observations may overcome this limitation, socioeconomic data that are often obtained from household surveys require much more effort to fully capture the cross links along the *Policy–Behavior, Behavior–Goal,* and *Goal–Policy* pathways and the evolving effects through time.

Taking biodiversity conservation as an example, conserving biodiversity requires the integration and synergy of many policy instruments in a "policy-mix" because environmental issues often have a mix of values and externalities that are addressed individually and separately in policy frameworks. Policy-mix is defined as "a combination of policy instruments which has evolved to influence the quantity and quality of biodiversity conservation and ecosystem service provision in public and private sectors" (Ring & Schröter-Schlaack, 2011). These mixes seek to build upon or complement existing regulatory, economic, and informational instruments across multiple objectives in addressing multiple and compounding factors contributing to environmental problems (Ring & Barton, 2015). Assessing the efficacy of PES programs to achieve biodiversity goals contains some level of uncertainty, where a PES policy instrument or a mix of instruments often amplifies similar goals from different angles (Ring & Schröter-Schlaack, 2011). The focus of policy-mix analysis is often to identify the most effective instruments and examine the role of individual policies in the mix, examining complementary combinations, counterproductive combinations, sequencing instrument combinations, and context-specific combinations to work apart from the networked character of implementing policies across levels and scales (Ring & Barton, 2015). Ring and Barton (2015) argue that overlap or redundancy of instruments can increase resilience. There is uncertainty about policy efficacy for biodiversity conservation, deeming this overlap as precautionary rather than inefficient (Barton et al., 2011; Ring & Barton, 2015; Ring & Schröter-Schlaack, 2011). However, these overlapping programs do not always have better effectiveness for conservation.

Furthermore, multiple issues involve complex systems of multi-level multi-actor governance where policies are created at different scales which address the same issues utilizing different forces of organizational power. For example, heavily subsidized agricultural or infrastructure investments may negatively impact biodiversity, leading to policy failure. The failure can be addressed through subsidy removal as an element of the policy-mix (Ring & Schröter-Schlaack, 2011). For example, Kubo et al. (2019) performed research highlighting the need to dissect the interactions between environmental, economic, and business development-related instruments into a multi-stakeholder policy-mix to mitigate anthropogenic disturbances on the national park conservation in Indonesia (Kubo et al., 2019). Gebara et al. (2019) addressed a "landscape approach" to tropical deforestation with diverse interactions between the natural and social space, commenting on the ineffectiveness of command-and-control policies, indicating that the failures were in ignoring cross-linkages in forestry, culture, conservation, and social development, highlighting an example from REDD+ in the Amazon (Gebara et al., 2019)

When applying a mix of policies across a landscape, where spatially explicit decisions are to be made, the term "policyscape" is applied (Barton et al., 2011, 2013). When considering the spatial distribution of policy on the landscape, there are many challenges in translating national and global level policy to the local land at subnational levels because of the uneven distribution of conservation priorities and economic sectors such as wildlife, forestry, and agriculture that may have higher costs than others (Ring & Barton, 2015). Evaluations for green initiatives—PES programs, for instance—have included prospective analyses using site selection models and post-assessment analyses using impact evaluation methods, which generally include spatially explicit features of the policy-mix, such as policy rights and financial incentives, to understand the cost–benefit scenarios of effectiveness (Barton et al., 2011). For example, studies investigating forests with high conservation value in Norway have shown that spatially overlapping policy instruments protected some types of forests between multiple policies, and other areas of high biodiversity value were not covered at all (Sverdrup-Thygeson et al., 2014). Another study investigated whether multi-level policies implemented by multi-stakeholder groups meant to address protecting forests of high deforestation risk were efficient or not, having operated in overlapping regions, finding that communities with large forests and low deforestation risks were actually enrolled, not areas of high deforestation risk, with some policies in the mix doing a better job than others in addressing the deforestation risk (Ezzine-de-Blas et al., 2016).

9.4 Outlook

Hidden spillover effects among concurrent green initiatives are prevalent worldwide despite geographic region, size, urban–rural gradient, developed–undeveloped spectrum, PES funders, payment stacking types, and so on. However, categorizing, detecting, and accounting for such spillover effects have a long way to go. The identified spillover effects in all selected cases are only a tiny subset of all potential ones, and except for the Fanjingshan and Tianma cases, all the spillover effects discussed herein were uncovered by researchers unintentionally. Uncovering these spillover effects remains difficult, if not impossible, in many other instances when spillover effects are not intentionally explored—as in the case of the Yucatán and Chiapas, where too many "no-data" records prohibited further exploration of the positive yet insignificant ($p = 0.14$) spillover effect.

Henceforth, our aim in this book is to raise awareness of the surge of concurrent green initiatives, paying attention to spillover effects that are often hidden or overlooked, even in many seemingly "successful" green initiatives worldwide. We call for robust scientific research on the magnitude, direction, and integrated effects of such hidden spillover effects, and the corresponding mechanisms and socio-ecological consequences. Such knowledge might provide crucial insights into many theoretical and practical issues in green initiative design, implementation, and evaluation. Also, such knowledge should be instrumental to maintaining many ecosystems and their vital life-support services (Díaz et al., 2019).

Global green initiatives face many theoretical and practical challenges—opportunities at the same time—in an era when "[t]he biosphere, upon which humanity as a whole depends, is being altered to an unparalleled degree across all spatial scales" (IPBES, 2019). By leveraging the widespread yet hidden spillover effects, governments and other relevant organizations can make these green initiatives more resilient to socioeconomic and biophysical crises such as the COVID-19 pandemic, effectively sustaining the environment.

References

Barton, D. N., Blumentrath, S., & Rusch, G. (2013). Policyscape: A spatially explicit evaluation of voluntary conservation in a policy mix for biodiversity conservation in Norway. *Society & Natural Resources*, *26*(10), 1185–1201. https://doi.org/10.1080/08941920.2013.799727

Barton, D. N., Primmer, E., Ring, I., Adamowicz, V., Blumentrath, S., & Rusch, G. (2011). Empirical analysis of policymixes in biodiversity conservation: A spatially explicit 'policyscape' approaches. *Instrument Mixes for Biodiversity Policies, Special Session*, 14–17.

Börner, J., Baylis, K., Corbera, E., Ezzine-de-Blas, D., Honey-Rosés, J., Persson, U. M., & Wunder, S. (2017). The effectiveness of payments for environmental services. *World Development, 96*, 359–374. https://doi.org/10.1016/j.worlddev.2017.03.020

Daily, G. C., & Matson, P. A. (2008). Ecosystem services: From theory to implementation. *Proceedings of the National Academy of Sciences of the United States of America, PNAS, 105*(28), 9455–9456. https://doi.org/10.1073/pnas.0804960105

Díaz, S., Settele, J., Brondízio, E. S., Ngo, H. T., Agard, J., Arneth, A., Balvanera, P., Brauman, K. A., Butchart, S. H. M., Chan, K. M. A., Garibaldi, L. A., Ichii, K., Liu, J., Subramanian, S. M., Midgley, G. F., Miloslavich, P., Molnár, Z., Obura, D., Pfaff, A., … & Zayas, C. N. (2019). Pervasive human-driven decline of life on Earth points to the need for transformative change. *Science, 366*(6471), eaax3100.

Ezzine-de-Blas, D., Dutilly, C., Lara-Pulido, J.-A., Velly, G. L., & Guevara-Sanginés, A. (2016). Payments for environmental services in a policymix: Spatial and temporal articulation in Mexico. *PLoS ONE, 11*(4), e0152514.

Gebara, M. F., Sills, E., May, P., & Forsyth, T. (2019). Deconstructing the policyscape for reducing deforestation in the Eastern Amazon: Practical insights for a landscape approach. *Environmental Policy and Governance, 29*(3), 185–197. https://doi.org/10.1002/eet.1846

Gren, I.-M., & Elofsson, K. (2017). Credit stacking in nutrient trading markets for the Baltic Sea. *Marine Policy, 79*, 1–7. https://doi.org/10.1016/j.marpol.2017.01.026

Hooper, D. U., Adair, E. C., Cardinale, B. J., Byrnes, J. E. K., Hungate, B. A., Matulich, K. L., Gonzalez, A., Duffy, J. E., Gamfeldt, L., & O'Connor, M. I. (2012). A global synthesis reveals biodiversity loss as a major driver of ecosystem change. *Nature, 486*, 105.

IPBES. (2019). *Summary for policymakers of the global assessment report on biodiversity and ecosystem services of the intergovernmental science-policy platform on biodiversity and ecosystem services*. Intergovernmental Science-Policy Platform on Biodiversity and Ecosystem Services. https://www.ipbes.net/news/ipbes-global-assessment-summary-policymakers-pdf

Kubo, H., Wibawanto, A., & Rossanda, D. (2019). Toward a policy mix in conservation governance: A case of Gunung Palung National Park, West Kalimantan, Indonesia. *Land Use Policy, 88*, 104108. https://doi.org/10.1016/j.landusepol.2019.104108

Liu, J., Hull, V., Batistella, M., DeFries, R., Dietz, T., Fu, F., Hertel, T. W., Izaurralde, R. C., Lambin, E. F., Li, S., Martinelli, L. A., McConnell, W., Moran, E. F., Naylor, R., Ouyang, Z., Polenske, K. R., Reenberg, A., Rocha, G. de M., Simmons, C. S., … & Zhu, C. (2013). Framing sustainability in a telecoupled world. *Ecology and Society, 18*(2), 26.

Millennium Ecosystem Assessment. (2005). *Ecosystems and human well-being*. Island Press.

Naeem, S., Ingram, J. C., Varga, A., Agardy, T., Barten, P., Bennett, G., Bloomgarden, E., Bremer, L. L., Burkill, P., Cattau, M., Ching, C., Colby, M., Cook, D. C., Costanza, R., DeClerck, F., Freund, C., Gartner, T., Goldman-Benner, R., Gunderson, J., … Wunder, S. (2015). Get the science right when paying for nature's services. *Science, 347*(6227), 1206. https://doi.org/10.1126/science.aaa1403

Program Evaluation Division. (2009). *Department of environment and natural resources mitigation credit determinations* [Special Report to the General Assembly Report Number 2009-04]. North Carolina General Assembly. https://www.ncleg.net/PED/Reports/documents/Wetlands/Wetland_Report.pdf

Quétier, F., Regnery, B., & Levrel, H. (2014). No net loss of biodiversity or paper offsets? A critical review of the French no net loss policy. *Environmental Science & Policy, 38,* 120–131. https://doi.org/10.1016/j.envsci.2013.11.009

Ring, I., & Barton, D. N. (2015). Economic instruments in policy mixes for biodiversity conservation and ecosystem governance. In J. Martinez-Alier (Ed.), *Handbook of ecological economics* (pp. 413–449). Edward Elgar Publishing. https://doi.org/10.4337/9781783471416.00021

Ring, I., & Schröter-Schlaack, C. (2011). *Instrument mixes for biodiversity policies: POLICYMIX report.* Leipzig: Helmholtz Centre for Environmental Research – UFZ. http://policymix.nina.no

Secretariat of the Convention on Biological Diversity. (2020). *Global Biodiversity Outlook 5.* https://www.cbd.int/gbo/gbo5/publication/gbo-5-en.pdf

Sverdrup-Thygeson, A., Søgaard, G., Rusch, G. M., & Barton, D. N. (2014). Spatial overlap between environmental policy instruments and areas of high conservation value in forest. *PLoS ONE, 9*(12), e115001. https://doi.org/10.1371/journal.pone.0115001

United States Department of Agriculture, & Farm Service Agency. (2019). *Conservation Reserve Program Statistics.* https://www.fsa.usda.gov/programs-and-services/conservation-programs/reports-and-statistics/conservation-reserve-program-statistics/index

USDA Farm Service Agency. (2016). *The conservation reserve program: 49th Signup results.* United States Department of Agriculture. https://www.fsa.usda.gov/Assets/USDA-FSA-Public/usdafiles/Conservation/PDF/SU49Book_State_final1.pdf

Waldron, A., Miller, D. C., Redding, D., Mooers, A., Kuhn, T. S., Nibbelink, N., Roberts, J. T., Tobias, J. A., & Gittleman, J. L. (2017). Reductions in global biodiversity loss predicted from conservation spending. *Nature, 551,* 364.

Wunder, S., Brouwer, R., Engel, S., Ezzine-de-Blas, D., Muradian, R., Pascual, U., & Pinto, R. (2018). From principles to practice in paying for nature's services. *Nature Sustainability, 1*(3), 145–150. https://doi.org/10.1038/s41893-018-0036-x

Yost, A., An, L., Bilsborrow, R., Shi, L., Chen, X., & Zhang, W. (2020). Linking concurrent payments for ecosystem services in a Chinese nature reserve. *Ecological Economics, 169,* 106509.

Index

For Product Safety Concerns and Information please contact our EU
representative GPSR@taylorandfrancis.com
Taylor & Francis Verlag GmbH, Kaufingerstraße 24, 80331 München, Germany